Can Tesla Save The World ?

I0478849

This book is dedicated to all the people in this world who wish
to preserve their home
: Planet Earth

Can Tesla Save The World ?

Index

Can Tesla Save The World ?

Can Tesla Save The World ?

(The generation of electricity: without the use of fossil fuels)

Introduction

Earth the final frontier.

All Star Trek fans will see the obvious flaw in this statement but unless this planet does not start to heed the warnings of world scientists, we may never have a chance of going into space to colonize other planets. We hear of plans to go to Mars but all these may be pipe dreams unless we clean up our act on our own planet first.

There has been much debate as to whether there is global warming or not, there has been speculation that the ozone layer will heal itself and that the oceans will somehow refresh themselves but there is much evidence that with the population growth of the human family, many resources will become scarce and thus we are going to meet challenges in the near future that our ancestors could hardly have dreamed about.

One problem that the world does own up to is that the demand for electricity will increase as technology gives us more innovations and as the former "third world" countries enter a new era of growth. Although there are many hydroelectric dams providing much of this power today the resources of the earth: coal, oil and natural gas play an important role in generating electricity. Nuclear power is also being used to a lesser extent due to the risk of radioactive contamination caused by either human error or natural disasters as has been seen in Japanese earthquake zones, such as Fukushima on 11 March 2011, or at the Chernobyl Nuclear Power Plant that occurred on 26 April 1986 in the town of Pripyat, in Ukraine. Another problem with nuclear energy is the radioactive waste disposal that is needed: a problem that has not been fully settled to everyone's satisfaction.

Can Tesla Save The World ?

Natural resource devices, such as solar panels, wind turbines, ocean currents, wave power and many others are making their contributions but usually on a limited basis for various reasons.

In the following pages you will read about a hydro plant that uses only two natural elements: air and water; in a way that neither of them will be contaminated or destroyed. You will learn that with the use of very simple siphons, tap valves and one-way check valves all the energy this world will need today and into its future will be readily available.

When topics that can be researched on the Internet are mentioned they are purposely not expanded upon, allowing the reader to do their own research into these topics. Some chapters will also give "You Tube" references to further add to this research.

Hopefully I have written the next few chapters in a way that even the novice experimenter will be able to run with these ideas and adapt them to their own situations. Please read on with an open mind and join the many on this planet who wish to leave a worthwhile legacy to their children and grand-children for many generations to come.

Can Tesla Save The World ?

The Tesla Story

Before we explore the inventions relating to the Air Electric system we need to understand why the title of this book refers to Nikola Tesla. Who was he? What contributions has Nikola Tesla made to our world so far and why could a man still be able to save our world today after he is dead?

Below you will read the story of Tesla and as you read it ask yourself how a man who came to be called the man of lightning and even labeled a mad scientist by cartoonists, be almost forgotten by the world. Ask yourself before you read his story, who invented radio, who discovered x-rays, who invented fluorescent lighting, who invented radio control, who invented the essentials of what we call today the star-wars program?

Nikola Tesla was born July 10, 1856 in a village named Smiljan, a subject of the Austro-Hungarian Empire. His parents were Serbian descendants. His father, Milutin, was an orthodox priest and his mother, Djuka, worked on a farm. Milutin was lovable but stern and an accomplished poet and writer, which writings Nikola loved to read as a boy. Djuka loved to create gadgets to help in the home and with the farm work. One invention Nikola remembered was a mechanical eggbeater that his mother made. He credited her with giving him his inventive mind. He was educated at home with his siblings when he was young but later attended a school in Carlstadt, Croatia, where he was an excellent student. One subject that he enjoyed was mathematics and he had the ability to work out integral calculus in his mind, for which feat his teachers accused him of cheating.

At some time in his youth he saw an engraving of Niagara Falls and told an uncle that one day he would harness the power of that waterfall by going to America. Since he loved mathematics and science, Nikola wanted to become an engineer but his father wanted him to enter the priesthood.

Can Tesla Save The World ?

When he was seventeen, Nikola contracted cholera, a potentially dangerous disease, which worried his parents greatly. Nikola had his father promise him that if he could overcome the ailment, his father would allow him to attend school to become an engineer. He survived and his father kept his promise, sending Nikola to the Austrian Polytechnic School in Graz and later to the University of Prague. Nikola began to study mechanics and electrical engineering, which included the study of a Gramme dynamo. (In 1873 Zenobe Theophile Gramme, a Belgium engineer, invented a motor/dynamo. It used two electromagnets and a commutator, to reverse the electric current direction.) When Nikola observed that the commutator's slip rings sparked as the electricity ran the apparatus, he remarked that the system was inefficient and should be replaced by a better method. His teacher replied that Nikola was suggesting a perpetual motion machine, which was impossible. His student, however, took up the challenge. Nikola decided that he would pursue his thoughts, knowing that somehow his idea would need to use alternating current. He would have been familiar with the work of Michael Faraday, that taught among other observations, that when a permanent bar magnet is pushed into a coil of insulated copper wire, an induced current would be produced. The current would generate in one direction when the magnet went into the coil, and then would reverse as the magnet was withdrawn. Nikola realized that his answer lay in the idea of induced currents.

At age twenty-four, Nikola moved to Budapest, where he worked at the Central Telephone Exchange. It was there that he came to the ideal solution as to how, his motor should work. He drew a diagram in the sand, with a stick, of his thoughts. Realizing that the electric motor would need to be a set of magnets on a rotor, spinning in front of a set of stationary magnets, Nikola perceived that these would have to be electromagnets, fed with alternating current. The stationary electromagnets would be arranged so that they would produce a north pole and then a south pole as the alternating current pulsed through them. The magnets on the rotor would also be

Can Tesla Save The World ?

electromagnets but they would be wound with wire in such a way that as the current was induced in one magnet it would produce a current in a corresponding magnet. This would cause the rotor to turn, one electromagnet attracting a rotor electromagnet, whilst another rotor electromagnet would be repulsed. Great thought had to go into how the voltage, current and magnetic fields were to be coordinated to achieve the right mix to produce the rotation, otherwise the motor would just hum backwards and forwards with no spinning motion. Nikola, being a mathematical genius, worked out how his induction motor would work. (The induction motor is also known as a squirrel cage, in that the alternating current will cause the north pole to be created slightly later than the one in the previous coil. The induced south pole in the moving rotor will thus chase the fixed north poles and thus, like a squirrel in a cage, is constantly chasing without catching up.) He also realized that the electromagnets in the stationary magnets were affected by the spinning of the moving electromagnets. With this thought, he concluded that if the motor could be started with alternating current and the rotor were kept moving by mechanical means, the induced current would now go to the stationary electromagnets and the motor would become a generator of alternating current.

As Nikola came to his conclusions, his work moved him to Strasbourg, Germany and Paris, France to help improve their direct current generation facilities. He tried in both countries to find investors for his new invention, the induction motor, but without success. Nikola decided that the only person that might listen to him was the, then, greatest electrical engineer, Thomas Alva Edison. This is when the six foot two inch, Nikola decided to emigrate from Europe to America and arrived in New York at the age of twenty-eight with only a few dollars to his name. He was not impressed with New York City but was determined to meet with Mr. Edison. He carried with him a letter from one, Charles Batchelor, an associate of Edison in Europe. Thomas Edison was quite successful at this point in time, having the backing of the financier J. P. Morgan. Nikola managed to have his meeting after Thomas had read the

Can Tesla Save The World ?

letter, telling Edison that Nikola Tesla was the greatest man Charles Batchelor had ever met, next to Edison himself.

Thomas Edison listened to Nikola's ideas for alternating current machines but viewed them as competition to his own direct current generators and motors. However, he was impressed with Nikola enough to hire the electrical engineer to improve the Edison direct current power stations. Nikola recorded that Thomas promised him $50,000. if he succeeded in accomplishing the improvements. However, this led to the first of many disappointments for Nikola, for although he completed the necessary work, Edison laughed off the $50,000. reward as an American joke, to which Nikola immediately resigned in disgust. He had put many long hours, with little sleep, into the project and felt betrayed by his hero.

Although Nikola ended up digging ditches in New York, news of his work came to the ears of others, including some businessmen who approached him to set out a proposal. They told him that they would be willing to put up the money to start a business, called the Tesla Electric Light Company, if he would develop an improved electric arc-light system. With his esteem partly restored, Nikola put his energy into the new company, inventing an arc lamp that was more efficient than its predecessors and a much better design. However, again disappointment struck as the backers took the profits from the business, leaving Nikola with some useless stock for his trouble.

Down again for a short time, another businessman, Mr. A. K. Brown of the Western Union Company approached Nikola proposing that he could pursue his lifelong dream to create an alternating current induction motor. Mr. Brown set him up in a small laboratory, near to Thomas Edison's offices, where Nikola designed and built both his motor and a means of transmitting such power, (ideas that we still use today.) Nikola filed for patents in 1887 for these ideas and was granted several, including one for an Electro Magneto Motor on May 1, 1888 and the Method of Electrical Transmission on June 25,

Can Tesla Save The World ?

1889. Patents followed for an Alternating Current Electro Magnetic Motor on August 5, 1890 and an Alternating Electric Current Generator on March 10, 1891. His patents for a system of Electrical Transmission of Power were granted between December 13, 1892 and December 26, 1893. It appeared that the world was ready for an electrical revolution. Nikola designed motors to use polyphase, as well as single-phase induction machines but it was his idea for power transmission that attracted another backer, by the name of George Westinghouse.

George, an inventor himself, of railroad air brakes, was also a Pittsburgh industrialist. He came to Nikola to offer him $60,000. to buy his patents with an agreement to pay royalties of $2.50 per horsepower of electrical capacity sold. (He actually paid him $5,000. and the balance in shares, in the newly formed Westinghouse Company, which meant that Nikola was at the mercy of stock again.) However, Nikola had confidence in his inventions and was grateful to George Westinghouse for allowing him to pursue his dreams.

Thomas Edison was now convinced that Nikola's alternating current ideas were in direct competition to his direct current method. He launched a campaign to try to discredit alternating current by trying to prove that it was a deadly and dangerous source of power.

However, in the midst of this electric battle, Chicago was to host the World Fair and wanted it to be an all-electric fair. The fair was called the Columbia Exposition as it celebrated the 400th anniversary of the date that Columbus discovered America. The Edison Company had been bought by the newly formed, General Electric Company and put in their bid for the fair, using direct current. George Westinghouse cut their bid in half to $500,000. proposing the more efficient, alternating current system. Again, Nikola went into action and by the opening of the World Fair on May 1, 1893, twelve one-thousand horsepower, alternating current generators started up, illuminating the fair with over one hundred thousand

Can Tesla Save The World ?

incandescent electric lights. It was estimated that twenty-seven million people attended the fair and were so impressed by it, that from that point onwards, the majority of electrical appliances and motors were powered by alternating current. One of those visitors was an Englishman named, Lord Kelvin; a famous British physicist. He had been an opponent of alternating current and, like Edison, felt the future was in direct current. However, this view changed dramatically when he visited the Columbian Exposition. When Lord Kelvin was called upon to head the International Niagara Falls Commission, he chose George Westinghouse as his first preference. His team approached George and received his agreement to use Niagara Falls as a means of generating alternating current electricity. The dream that Nikola had as a youth was coming true.

With the help of some of the wealthiest men in America and Europe, including J. P. Morgan, the five-year project began. These years were plagued with doubts of investors and the stress of the workers, as the equipment was constructed. Nikola had envisioned this generation plant in his mind for years and as with many inventors, Nikola could see, almost as a vison, the details of his invention, down to the last screw. He had experimented mentally, solved problems mentally and had seen the final results mentally, long before it was a physical reality. When the switch was pulled at midnight, November 16, 1896, the operation went without a flaw and Buffalo, New York received its first five thousand horsepower of electricity, which went to the railroad. The demand for an additional five thousand horsepower came soon after and within the next few years ten generators were providing the power to light New York City and its other needs.

However, because of litigation suits between Westinghouse and General Electric the money reserves were greatly drained. J. P. Morgan attempted to take over the electric industry and George Westinghouse appeared to be on the brink of ruin. He went to Nikola and asked him to give the Westinghouse Company a release from the $2.50 per

Can Tesla Save The World ?

horsepower agreement. Nikola tore up the contract, saving George Westinghouse but creating a further history of financial woes for himself.

At this point we shall try to put ourselves in the place of Nikola. Since his days at school, he had wanted to build a motor and generator that was more efficient than the Gramme dynamo: he had succeeded. From his youth, he also wanted to harness the power of Niagara Falls: he had succeeded. Though he had not become rich from his inventions and had been cheated by so-called business associates, it appears that he still had a friend in Mr. A. K. Brown, for the laboratory on Grand Street, New York was still available to him. Yet where was he to go from here? It appears that one thing Nikola was always thinking about was how to make anything that was electrical more efficient. Two ideas that seemed to occupy Nikola's mind were that the incandescent light bulb had only five percent efficiency as far as a light source and that the wires carrying the electrical transmission of his alternating current were cumbersome and needed more efficiency. From his studies of such men as James Clark Maxwell and Heinrich Rudolf Hertz, Nikola knew that the accepted theory of light was that it was made up of electromagnetic radiation, vibrating at very high frequencies. Also, that these electromagnetic waves could travel far out into space. Possibly, Nikola concluded that if he could reproduce the high frequencies of light, he could solve both problems: a better source of light and electricity being transmitted through space.

At first, he experimented with improving his induction generator to increase its speed and therefore the cycles per second. Mathematically, he knew that the frequencies he would need would have to be in thousands of cycles per second but he quickly realized that his generators could not go from sixty cycles per second to what he wanted, without the machines failing. What he needed was a way of transforming the regular alternating current to the higher frequency. It was well known that a circular core of metal, wound on one side with extra-fine insulated copper-wire, wound many times, and

Can Tesla Save The World ?

the other side with a thicker insulated copper-wire, would make a transformer. The result would be that the amperage on this fine wire side would be lowered greatly and a corresponding increase in the voltage would occur, but how could this increase the number of cycles?

Nikola was also familiar with the Leyden jar, which acts as a capacitor. He realized that as the increased voltage built up in the jar, it would eventually discharge across a spark gap. The work of Hertz had taught Nikola that such a spark was a burst of high frequency electromagnetic waves, but going in one direction. He needed to reverse the direction many thousands of times per second to achieve his vision.

He began by constructing a transformer that had no iron core. Instead, knowing that ionized air would be conductive, he made a core of heavy copper wire, into which he placed a core of fine insulated copper wire, wound many times. When the primary coil, (the heavier copper wire), was connected to an alternating current source, it induced a current in the secondary coil, producing a high voltage output.

Nikola than connected a Leyden jar to the primary coil so that it would act as a capacitor, storing electricity that would be discharged across a spark gap, also connected to the primary coil. His idea took into account that a capacitor allows current to build quicker than the voltage, whereas in a coil inductor, the voltage builds quicker than the current. He calculated that by making the capacitor and inductor a certain size, the voltage and current would never be able to obtain a stable position and therefore oscillation would result. The spark gap allowed discharges of electricity as the air between the space became conductive. The air gap then became an insulator, allowing the voltage and current to build again but in the opposite direction. Again, the air in the spark gap was conductive and a spark in the opposite direction occurred. This discharge and recharge occurred many thousand times per second. Added to this, the secondary coil was so designed

Can Tesla Save The World ?

that the voltage peak became a wave that would appear to be standing still.

This brilliant invention, (called the Tesla Coil), meant that Nikola could achieve many applications with this relatively simple device. He knew that certain gases became incandescent when electricity passes through them and found that his high frequency electricity caused light emissions in phosphorescent substances. In 1890 he succeeded in lighting phosphorescent tubes at a distance with no wires attached to them, making him sure that both of his ideas would work: a more efficient lighting system and that electrical power could be transmitted without wires. (During these experiments, he also developed the first neon and fluorescent lighting, as well as taking x-ray photographs.)

In the next few years he continued to develop his wireless ideas, though several patents were granted to him concerning the transmission of electrical power, a reciprocating engine, an electrical railway system, an electrical meter, a steam engine and an electrical condenser. By 1895 Nikola had perfected his coil to the point where he would receive and send radio waves from one coil to another, when he turned the resonance of one to be in synchronization with the other. He was about to transmit radio signals from his laboratory to a point fifty miles away when a fire, in the building, destroyed his apparatus.

Whilst he wrestled with this disaster, in England a young Italian, named Guglielmo Marconi, was working on wireless telegraphy. Marconi managed to file an English patent of a simple system in 1896, a year before Nikola filed for his basic patent in the USA.

In 1898, Nikola demonstrated radio to the public at an Electrical Exhibition in Madison Square Gardens. He used the apparatus, for which he was granted a patent in that year, to make a radio-controlled boat. The vessel was a metal boat, about four feet long, having several batteries on board to drive a propeller, rudder and some lights. Nikola also invented a

Can Tesla Save The World ?

switching device, by partly filling a thin canister with some metal oxide. When the radio waves reached the canister, the oxide became conductive and allowed a current to flow through it to operate a disk. As the disk turned, it made contact to activate the propeller, rudder and lights, sending the boat to the left with one signal and the right with another. Between the signals, a set of gears and springs turned the canister over, thus making the oxide non-conductive until the next radio signal was received. As with many of Nikola's ideas, the public did not appreciate the genius of what they saw, for they were experiencing the beginnings of robotics and the infancy of the remote-control system, used today in the space program.

Nikola's patent for radio was granted on March 20,1900 and the patent for an apparatus for transmission of electrical energy on May 15, 1900. Although
Marconi was busy in England creating the Marconi Wireless Telegraph Company and attracting investors, Nikola was working on his ideas. He realized that when air is thinner it will become more conductive and thus turned his attention to using the upper regions of the atmosphere to transmit radio waves.

Nikola's friend, L. E. Curtis and a former backer of the Niagara Falls project, Colonel John Jacob Astor, came to his aid, by supplying land in Colorado Springs and money to set up a laboratory at Pike's Peak. The area was famous for lightning storms and Nikola studied the effects of lightning strikes on the earth. His conclusions were that the earth acted as a giant conductor and that the lightning caused waves that traveled far beneath the earth, probably around the outer crust. He theorized that it should be possible to transmit electrical energy from one point to another as long as there was a way of making artificial lightning; of the magnitude seen in real storms. With the help of his associates, Nikola built a wooden tower that was eighty feet high. On top of this was placed a one hundred and forty-two-foot metal mast, capped by a large copper ball. In a building at the base of the tower there was a

Can Tesla Save The World ?

large Tesla coil that emitted powerful impulses. When Nikola was content that all the apparatus was connected correctly, he waited until nightfall to try his experiment. An assistant manned the Tesla coil and at a signal from Nikola, threw the switch for one second. The secondary coil produced a blue corona in the air around it, which told Nikola all was ready. He then instructed his assistant to throw the switch again but this time to leave the Tesla coil going until Nikola told him to switch it off. This time large areas of the blue light ran up and down the mast until bolts of lightning, measuring more than one hundred feet, shot out from the copper ball. Before the experiment was finished, Nikola succeeded in burning out the dynamo at the El Paso Electric Company and putting the closest city into total blackness.

However, undaunted by their distain, Nikola pressed on. He made several experiments over the next nine months, at the facility at Pike's Peak. Although it is not known if the experiments satisfied all his expectations, Nikola returned to New York and wrote an article in the "Century Magazine", outlining several of his findings. Among them was a method of using the sun's radiation as a form of energy, by using an antenna. He outlined machines that would make war impossible, (or so he believed), and suggested that electrical energy could be used to control weather patterns. He also proposed a worldwide system of wireless transmission. Many, including Nikola's former backer, J. P. Morgan, read the article. Mr. Morgan gave Nikola $150,000. to build a radio tower, thinking that he could tap into the new world of wireless communication. However, Nikola was not so much interested in conveying messages as he was in demonstrating to the world that electrical power transmission was possible without the cumbersome wires, which we have come to accept.

Nikola chose the cliffs of Long Island to build his new tower, known as the Wardenclyffe project. Stanford White, designer of Madison Square Gardens, was hired to design the laboratory, which was completed. The tower was not completed, although it was built to the point where there was a

Can Tesla Save The World ?

one hundred and eighty-seven-feet edifice, topped by a fifty-five ton, steel sphere. In the base was a spiral staircase that extended one hundred and twenty feet, via a shaft, into the ground. Below this were sixteen iron rods that went three hundred feet into the earth.

All appeared to be going well until more funds were needed. J. P. Morgan suddenly became reluctant to fund the project and then the stock market crashed, (1901), causing creditors to start pushing their claims on the venture. To add another dilemma, Marconi succeeded on December 12th 1901 in transmitting the Morse code letter "S" across the Atlantic Ocean from England to Newfoundland, Canada. At first Nikola dismissed the news, knowing that Marconi was using seventeen of his patents, but by 1905, when J. P. Morgan finally made it impossible for the Wardenclyffe project to continue, Nikola was devastated. Marconi was gaining ground, in that the US patent office granted him two patents that had been acknowledged as belonging to Nikola. When the Nobel Prize went to Guglielmo Marconi a few years later, Nikola was angry and crushed, mentally as well as materially.

Crushed but not defeated, Nikola produced a new type of turbine in 1912. He took a series of disks and placed them close together on a common shaft, in such a way that when steam or another vapor was fed to them, the disks rotated swiftly. However, it was found that the disks became distorted, due to the heat created, and so the idea was scrapped. The turbine appeared to have been a variance from Nikola's past ideas but in 1914 he described a method of using high-frequency radio waves to detect enemy ships, by detecting their metal hulls on a fluorescent screen. He also described rockets that were remotely controlled to become bombs. Both ideas were far in advance of their time, yet radar and V-bombs became players in the Second World War of 1939 to 1945.

On January 3, 1928, Nikola, then seventy-two years old, received his last patent entitled, "The Method and Apparatus for Ariel Transportation". Although, in theory, the invention

Can Tesla Save The World ?

resembled a cross between a helicopter and an airplane, (Nikola did not have the money to build the prototype), the concept closely resembled vertical lift and take-off planes of today. The US patent office had recorded one hundred and eleven utility patents, one reissued patent and two utility patent corrections in the name of Nikola Tesla, by this time: no small feat for any man. Yet his mind did not stop working and in 1934 the New York Times declared that Nikola had invented an invention that, "would send concentrated beams of particles through the free air, of such tremendous energy that they will bring down a fleet of 10,000 enemy planes at a distance of 250 miles." By 1937 Nikola had decided to share his idea with the world and sent details to the USA, England, Canada and France with little or no response. He also shared his idea with the (then) Soviet Union and Yugoslavia. He titled the paper, "The New Art of Projecting Concentrated Non-Dispersive Energy Through Natural Media": apparently describing what we know today as a charged particle beam device. Russia appeared to be the only one interested and sent Nikola a cheque for $25,000., after they had finished their first stage of testing the unit.

Nikola had hoped that not only would his particle beam idea bring peace to the world but that it would also be the answer to his ultimate-goal: to transmit electrical power over long distances without the aid of wires or hydro lines.

He died January 7, 1943 at the age of 86, after being responsible for so many brilliant and far-seeing ideas and inventions but his name was almost obliterated from the scientific world. Only recently has there been a resurgence of interest in this great man. As time passes some of Nikola's inventions are being revisited and are being adopted into our modern world. Below you will understand then why this book bears his name in the title.

You Tube – **The Life of Nikola Tesla**
https://www.youtube.com/watch?v=98QwPO1b5j4

Can Tesla Save The World ?

Overview of the System

The major elements that will be used will be water, converted to distilled water, and compressed air: both of which will be used in such a way that after use they will be recycled or allowed to return to their natural state.

Although some of the methods to generate electricity are not new, we will introduce a new kind of pump that uses the weight of water as its motive power and a new method of using electrolysis to heat water to steam, but not for steam's usual purposes in electrical generation.

It will be explained that this method is in no way perpetual motion and the use of supplementary power from solar and wind energy machines will also be utilized.

Water for the initial project will come from nearby lakes or rivers, supplemented with rainwater and snow.

The system will be initiated by a mother-plant; that will provide compressed air. Further satellite plants will generate electricity for their local areas. Because the compressed air will be transported by underground piping there will no need for the usual power lines and their massive towers that are seen everywhere in our modern world.

In rural areas and some business' there will be a different type of satellite station.

Historically this method of generating motive power was used in Paris, France. The Austrian engineer, Viktor Popp constructed a compressed air system in 1888 of 1,500 KW, building to a volume of 18,000 KW by 1891. This was sent through over four miles of main artery piping and thirty miles of sub arteries, primarily using the Paris sewer system, to factories, shops and homes to power the many uses of compressed air machines that were invented at the time, including electrical generators. Originally Popp used the River Seine to send water to a coal fired steam-generator that created the compressed air. The air pressure appeared to be around 80 psi that was piped directly into the main arteries of his system. The compressed air system was used in Paris up until 1994 when it was terminated. (For a more detailed explanation of Popp's system, refer to the Internet.)

The following chapters will introduce the various parts of the Air Electric system; which will eventually come together to show the whole process of generating electrical power without the use of fossil fuels.

Can Tesla Save The World ?

Siphons

Historically siphons have been around for a long time. One ancient mention of them came from the famous Grecian mathematician and philosopher Pythagorus. A wine merchant became upset that although he allowed his workers to drink his wine, they abused the system by drinking too much. The merchant commissioned Pythagorus to create a cup that his workers could use but somehow ration the amount of wine in it. Pythagorus then invented what has become known as the Pythagorus cup or the Greedy cup. Essentially it is a normal goblet with a column in the middle of it. Within the column is a tube that allows the wine to go to a certain level but if the cup is filled beyond that level the siphon kicks in and the excess wine flows down the column and onto the greedy person's lap, if he is sitting to drink. The merchant was apparently happy and the cup is still produced in Greece but as a novelty joke item. (See Fig. 1.)

The siphon was also known by the Romans but since we see aqueducts carrying water but no siphons it is likely they did not use their knowledge. At first sight one may wonder why a siphon would be used when to let water spill over the lip of a container would appear to do the same function. However, when we understand why a siphon works the use of the siphon will become clearer.

The first factor that is relevant to both siphons and liquid running over a lip is gravity. The liquid flowing out of the long side of a siphon acts because gravity is pulling the liquid downwards.

The second factor is air pressure in that the pressure of air at sea level is about 15 pounds per square inch which is pressing upon both the surface of the liquid running over the lip and the water above the siphon. The air pressure is therefore forcing the liquid up the short side of the siphon. Interestingly the air pressure allows the short side of the

Can Tesla Save The World ?

siphon to be up to thirty feet in the air at sea level and the siphon will still function so long as the longer side of the siphon is below the short side of the siphon.

The third factor that relates to the siphon but not the liquid going over a lip is a vacuum. When the water is drawn out of the long side of the siphon by gravity it creates a vacuum in the bend of the siphon. This combined with the air pressure pushes or sucks the water in the short side to fill the vacuum and thus allows the liquid to pass over the bend onto the longer side of the siphon. This also is a factor as to why the siphon can work up to thirty feet but not above this height.

The fourth factor relating to the siphon but not the liquid going over a lip is the volume of liquid on the long side of the siphon. Since the longer side of the siphon can be measured then the volume of liquid within the siphon can also be measured. This factor is important when it comes to the electric generator as to create a frequency of 60 cycles per minute (in North America) or 50 cycles per minute (in Europe). In conventional generators, a turbine known as a Frances turbine has a method of controlling the water coming into it to achieve the needed frequencies but as we will be using Tesla turbines (described later) the siphon will allow us to measure the water entering the Tesla turbines and thus the frequency can be controlled.

As will be seen from the air electric system below the Pythagorus cup siphon will be the method used for passing water from one level to another.

You Tube – **How a Siphon Works**

https://www.youtube.com/watch?v=CZmP0vsRBZ8

Can Tesla Save The World ?

Siphons

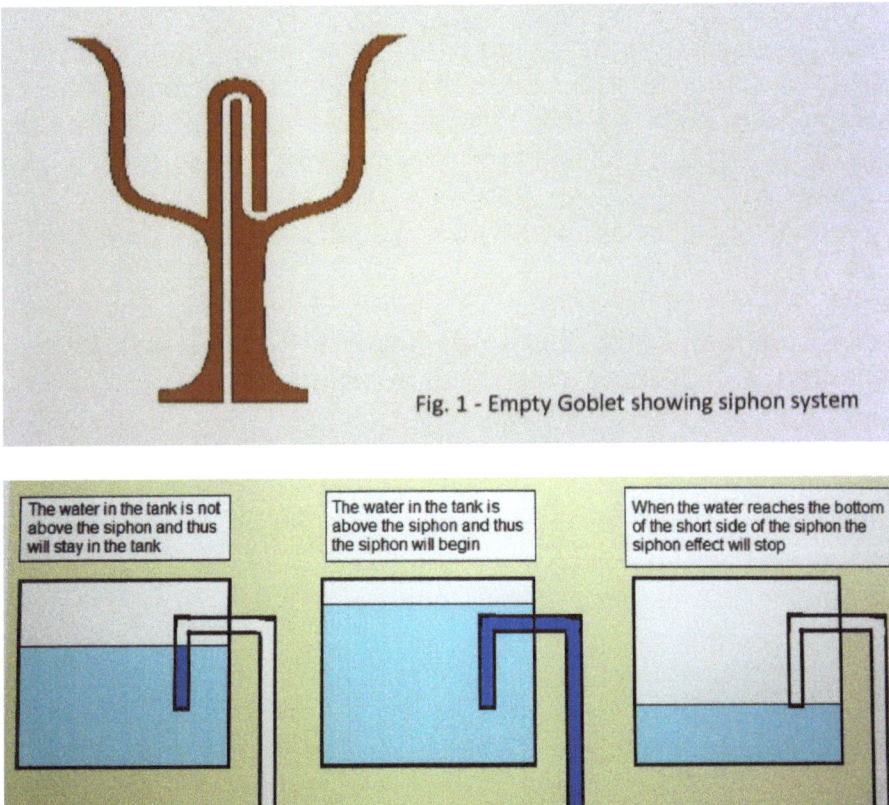

Fig. 1 - Empty Goblet showing siphon system

The water in the tank is not above the siphon and thus will stay in the tank	The water in the tank is above the siphon and thus the siphon will begin	When the water reaches the bottom of the short side of the siphon the siphon effect will stop

Fig. 2

Can Tesla Save The World ?
Water Pump

There have been many versions of the water pump: from the traditional hand pump up through large industrial electrical and steam driven pumps but as far as I can ascertain there has not been a water pump utilizing the weight of the water itself.

The weight of one gallon of water at sea level is estimated at 8.34 lbs. This means that a system can be created to hold a certain volume of water to drive a piston downwards and a further method to raise the piston upwards creating a water pump that can become a practical appliance.

Fig. 1 illustrates the overall action that is envisaged to make this pump work. Basically a source of water, shown as tank A, is filled with water from an outside source, (river, lake, stream, etc.), such that when it reaches the top of the tank a siphon (B) begins to empty tank A into tank C.

Tank C is connected to a movable piston (G) such that as the tank fills with water the weight pushes the piston downwards. Tank C also has a siphon (D) that comes into play when tank C is full.

This siphon will cause the water in tank C to empty into tank J where there is a float (E) that is connected to tank C. As the water enters in tank J the float will rise and thus the now empty tank C will rise with the float to its original position ready to accept its next influx of water from tank A.

Note that when float E reaches the top of tank J, siphon F will pass the water in tank J to tank H, which in turn will be sucked into pump "I" when the piston G is being moved up by float E.

Piston G, which is connected to tank C, will therefore act as a water pump and be able to pump water from tank H to a greater height than that of tank A.

Can Tesla Save The World ?

The system will not be 100% efficient and so any excess water will flow from tank H into a pipe that may operate another water pump system or return its contents to the original stream or river, but at a lower level downstream.

The system will also be slow but as this is not a factor in our use for this water pump it will not be a deterrent.

Fig. 2 shows tank A in better detail and shows that there will be several siphons coming from the tank allowing the water to flow quite quickly into tank C.

Fig. 3 shows tank C. Note that there are also more than one siphon coming from this also feeding more than one tank J below. This also means that instead of one float there are several floats at work to raise tank C.

Fig. 4 shows the working of tank J and how it will enter into the tank where the actual water pumps are situated. Note that the overflow siphons will send the excess water to tank K, which will act as a tank A or else to piping which will take the excess water to a stream or river at a lower level.

The use of this pump will be shown later, when the whole operation of the air compression plant (mother plant) will be discussed.

Can Tesla Save The World ?

Water Pump

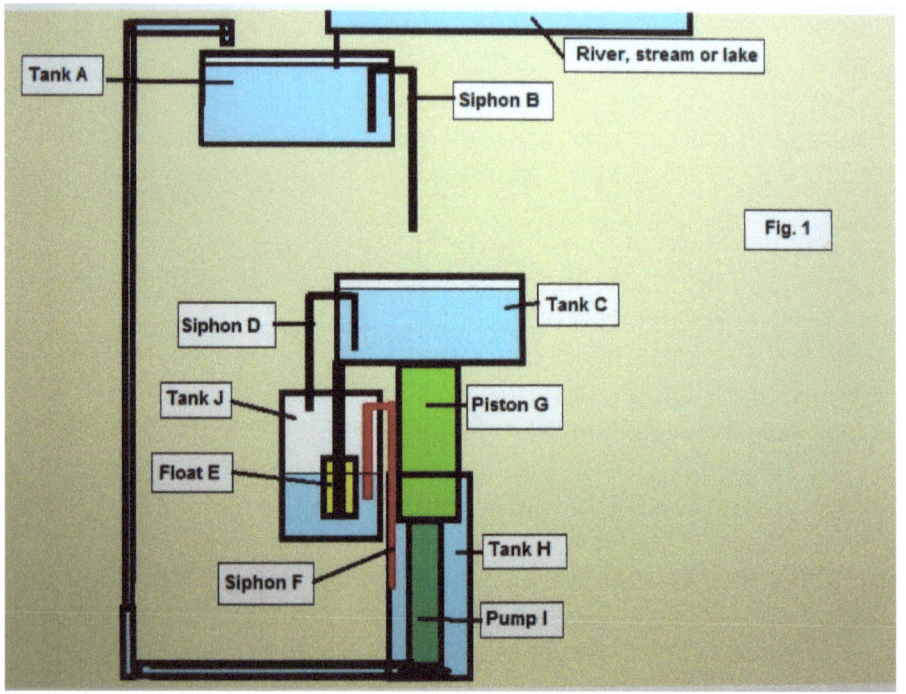

Can Tesla Save The World ?

Water Pump

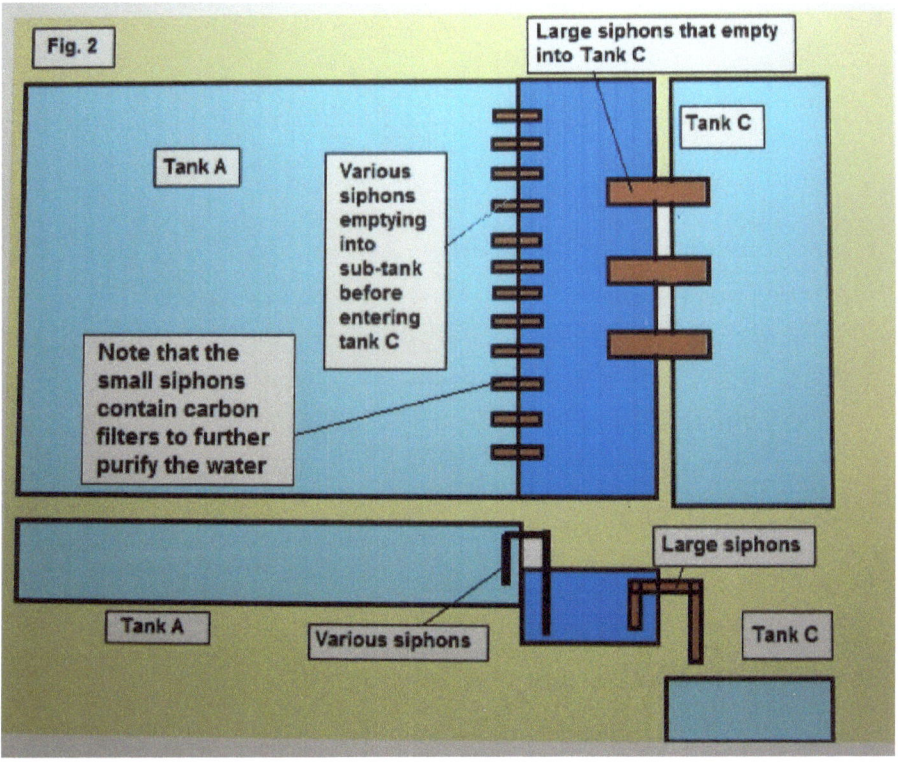

Can Tesla Save The World ?

Water Pump

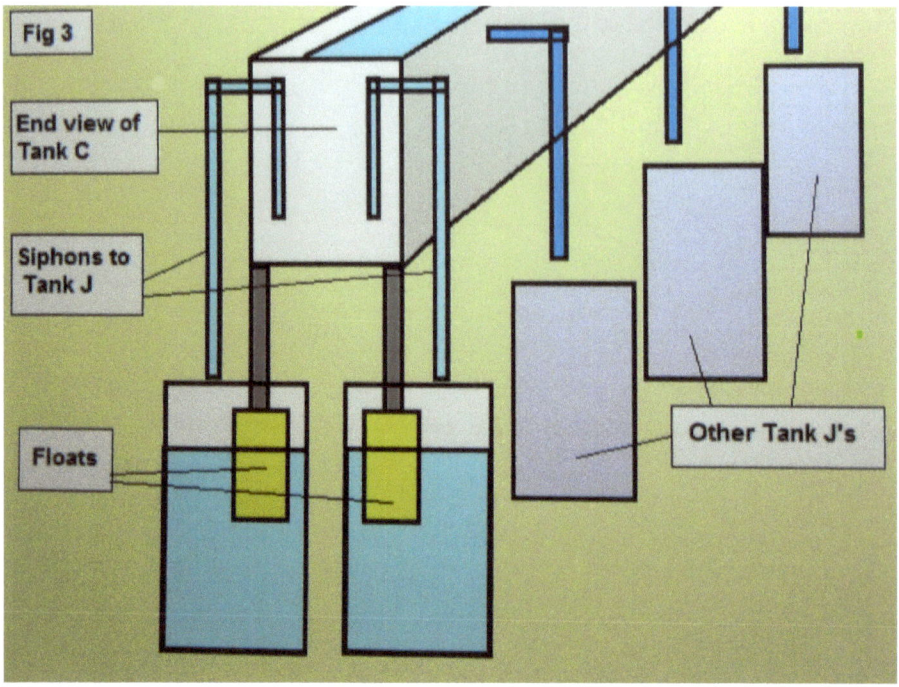

Can Tesla Save The World ?

Water Pump

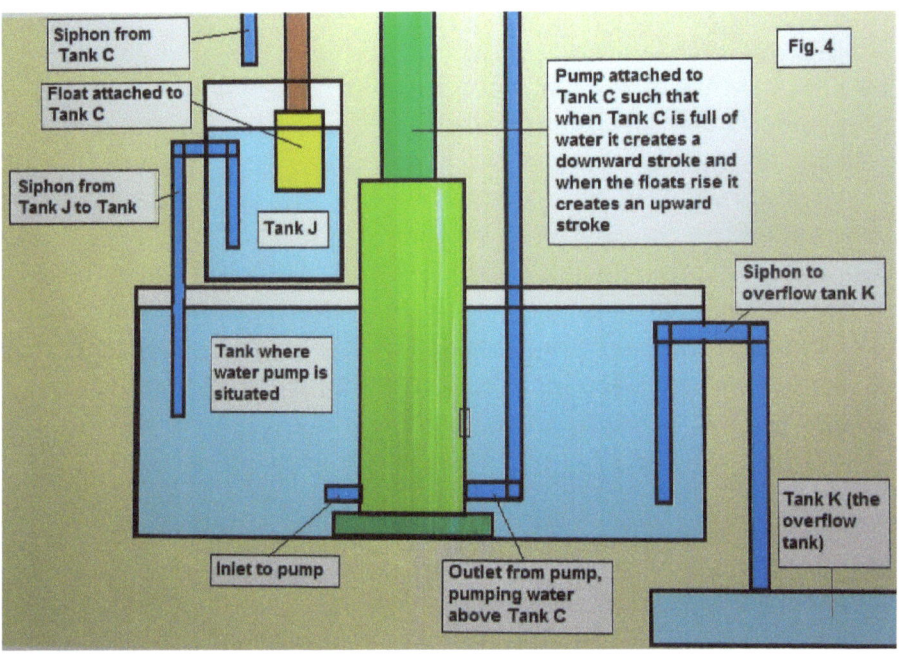

Siphon from Tank C

Float attached to Tank C

Siphon from Tank J to Tank

Tank J

Pump attached to Tank C such that when Tank C is full of water it creates a downward stroke and when the floats rise it creates an upward stroke

Fig. 4

Siphon to overflow tank K

Tank where water pump is situated

Inlet to pump

Outlet from pump, pumping water above Tank C

Tank K (the overflow tank)

Can Tesla Save The World ?

Trompe

The Trompe is definitely not a new invention. In reality it may have been around from as early as 1560 when it was used as the means of generating compressed air for the Catalan forge in Spain. As steam became the dominant energy source of the Industrial Revolution and later electricity and oil, the use and purposes of the trompe were almost forgotten. One large trompe is still in existence near to Cobalt, Ontario, Canada, on the Montreal River: it is the Ragged Chutes facility, being 351 feet deep and nine feet in diameter; it provided compressed air to the nearby mines for ventilation and power for mining equipment. Designed by Charles Taylor it was built in 1910 and though not used today it still stands as an engineering marvel to this day.

The trompe is a very simple device as illustrated in Fig. 1 and 2, where a river with a head of ten feet or more, is allowed to flow into a funnel like arrangement. The funnel is a structure that could be made of aluminum that has several holes running through it that makes it look like a giant sieve. The head of water, thus running into a trough where the funnel is situated, draws air through the openings created by the holes. Although the air has a tendency to rise, the velocity of the water and its weight create the Venturi effect in the constriction and thus the air is compressed as it descends down the pipe of the trompe. The temperature of the water determines the temperature of the compressed air. Any heat from such compression will be absorbed by the surrounding water, which means that the resultant compressed air will be very clean and as it becomes isothermal compression, the air will be cool.

The compression of the trapped air will be 14 psi for every 25 feet drop in the water level and so as this project will have a fall of 150 feet for each trompe the resultant compressed air will be about 84 psi. It is expected that each trompe will be about six feet in diameter.

Can Tesla Save The World ?

Fig. 1 shows that each trompe drops ten feet from the top of the funnel to its outlet. This means that if the drop of the first reservoir is also ten feet there is a distance or drop of forty feet from the top of the initial reservoir to the top of the over flow siphon.

Thus, as can be seen in Fig. 1, although the water in the initial tank flows into the first trompe the outlet at the third trompe will allow the water to overflow via a siphon to a fall of another 20 feet to the Tesla turbines. Therefore, from the top of the initial tank to the top of the Tesla turbine overflow tank there will be a distance of about eighty feet. (This fact will be important to remember when the steam boiler pipes are discussed.)

Fig. 2 shows the trompe in greater detail and how it is intended to be fed by a reservoir encircling the funnel using siphons to bring the water into the venturi effect opening.

Thus the trompe becomes the major source of the compressed air and the main purpose of the mother plant, but now the problem arises in that although the water has descended 80 feet how do we raise the same water to keep the operation going?

This will be explained in the following three chapters:

You Tube – **Bill Mollison explains a Trompe**

https://www.youtube.com/watch?v=SScpJMsCm9c)

Can Tesla Save The World ?

Trompe

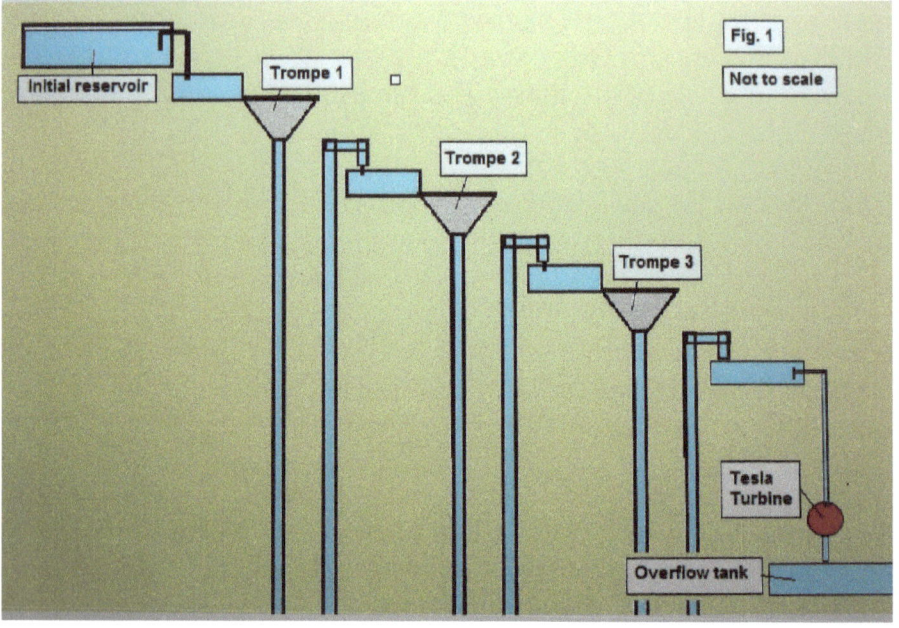

Fig. 1

Not to scale

Initial reservoir

Trompe 1

Trompe 2

Trompe 3

Tesla Turbine

Overflow tank

Can Tesla Save The World ?

Trompe

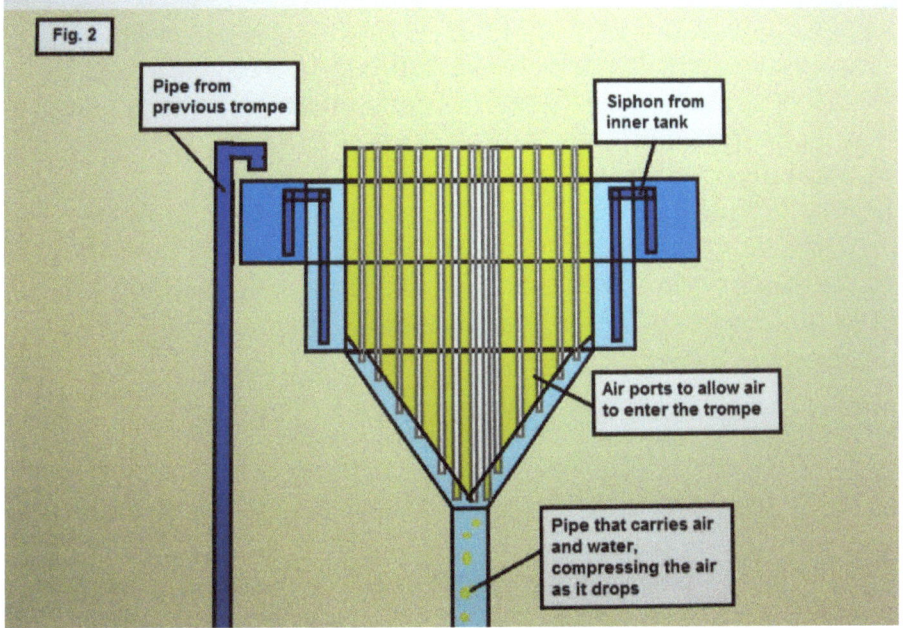

Fig. 2

Pipe from previous trompe

Siphon from inner tank

Air ports to allow air to enter the trompe

Pipe that carries air and water, compressing the air as it drops

Can Tesla Save The World ?

Tesla Turbine

Nikola Tesla, the inventor of the alternating pulsed electric motor and generator, which allowed us to have our modern electrical system, also invented many other ground breaking inventions including one that he is reported to have labeled his greatest invention: The Tesla Turbine. Nikola looked at the modern turbines of his day and realized that they were not very efficient, mainly because of the energy loss due to the steam, water or gas driving them having to change direction several times to produce the necessary torque to turn the turbine.

Nikola was familiar with the Bernoulli principle that basically relates to gas flowing over a surface creating a skin as it were so that it would appear that the gas adheres to the surface. This principle is well known in the aviation field but Tesla looked at it as a means of turning a turbine. Basically he took thin stainless steel plates, circular in shape and cut three holes near to the centre of the disk, (see Fig.1) and joined ten of these together using bolts and washers. He found that when he passed either compressed air or other gas, or steam or water on one side of the turbine the Bernoulli principle meant that the adhesion would be set up so that the turbine would begin to spin allowing the gas, water or steam to exit via the three holes in the disks. He also found that this action was very efficient and claimed an 85-90 % efficiency rating to his turbine. He also discovered that by having the washers positioned near the edge of the turbine he could create the needed torque. He then connected his turbine to a generator and thus produced the power to generate electricity, (he used steam to run the turbine). However, due to the high velocity of the turbine he found that the disks began to buckle after a while and therefore it has not been used as a steam generator in today's world, though modern materials like carbon fiber, graphene and other new improved metals would probably make it work.

Can Tesla Save The World ?

The Tesla turbine for our project would also utilize metal and plastic combinations to create a very effective water powered generator. In the air electric system, a twenty-foot head of water from the trompe siphon outlet will power the turbine. Since the disks will be 10-12 inches in diameter and arranged in ten disk increments, the siphon will not have to be too large in its diameter. The torque needed will be such as to turn the homopolar generators (later described) and thus not a lot of power will be necessary, particularly as the intent will be to have the turbines in series or tandem so that several sections of homopolar generators will be powered at once.

In Fig.1 it illustrates the disk to be used, but the history and application of the Tesla Turbine may be found on the Internet as well as shown in You Tube videos.

Fig. 2 shows how the turbine will be connected to the homopolar generators without the water from the turbine coming in contact with the generator. The exit of the turbine allows the water to enter the reservoir for the pumping system that will feed the trompe in the mother plant.

It is not known, as yet, what the efficiency of the turbines will be but it will definitely be higher than conventional turbines of today.

Fig.3 shows the arrangement of the siphons and the Tesla disks so that several siphons will power each set of homopolar generators. Note that the generators are connected in series, allowing for more power to go to the electrolysis process.

Although the Tesla turbines and homopolar generators will provide electricity, it has been stated earlier that the air electric system is not a perpetual motion machine and there will be additional electrical power provided by batteries, powered by windmills and solar panels above ground of the mother plant.

Can Tesla Save The World ?

However, the Tesla turbines will be a powerful source of electrical energy not only for the electrolysis system but also in other areas of the air electric operation.

You Tube -**The Tesla Turbine and How it Works**

https://www.youtube.com/watch?v=mrnul6ixX90

Tesla Turbine

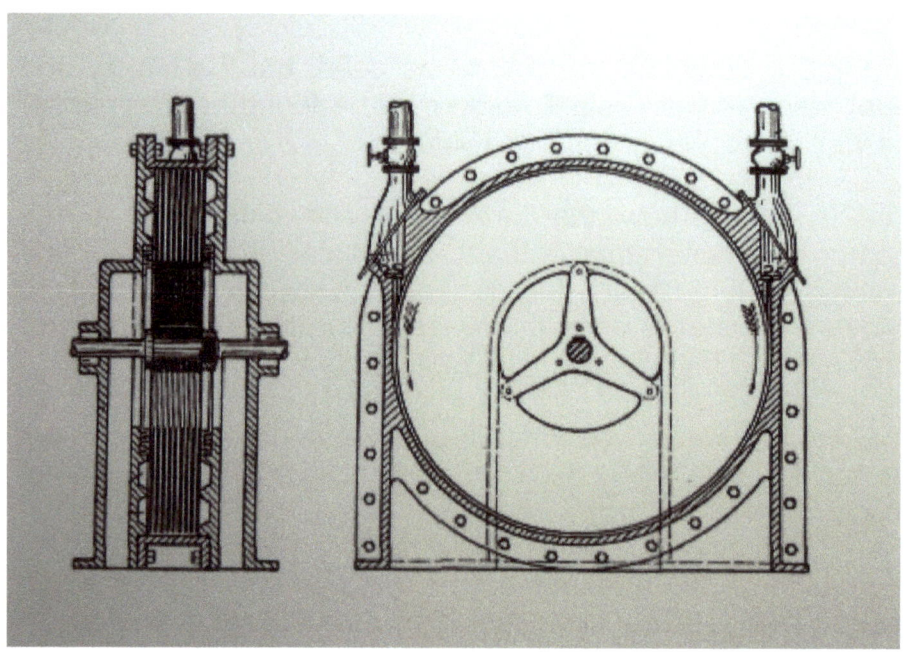

Can Tesla Save The World ?

Tesla Turbine

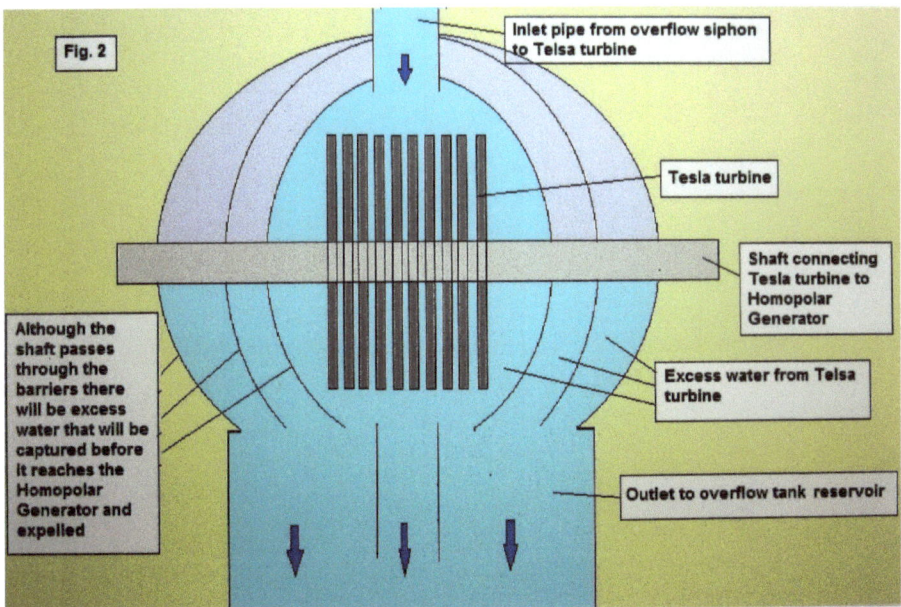

Fig. 2

Inlet pipe from overflow siphon to Telsa turbine

Tesla turbine

Shaft connecting Tesla turbine to Homopolar Generator

Although the shaft passes through the barriers there will be excess water that will be captured before it reaches the Homopolar Generator and expelled

Excess water from Telsa turbine

Outlet to overflow tank reservoir

Can Tesla Save The World ?

Tesla Turbine

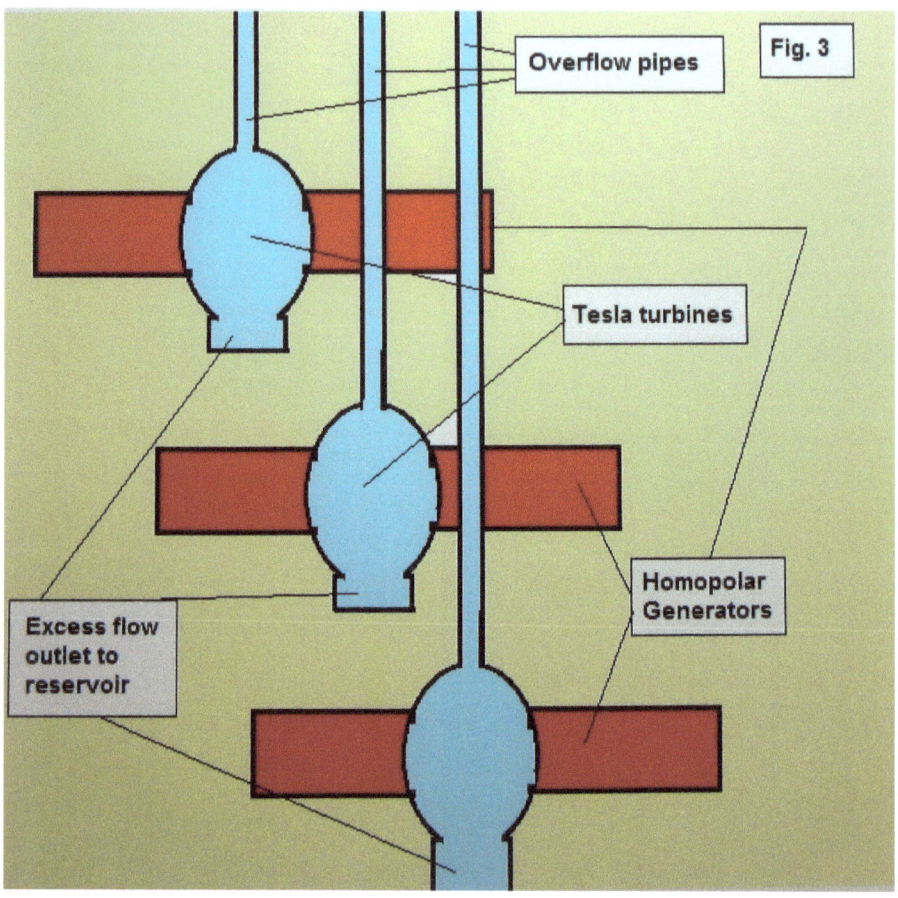

Overflow pipes

Fig. 3

Tesla turbines

Homopolar Generators

Excess flow outlet to reservoir

Can Tesla Save The World ?

Tesla Homopolar Generator

The famous electrical pioneer and scientist, Michael Faraday, discovered that magnetism and electricity were connected when he constructed a device in which a copper disk with a copper spindle was turned by hand within the poles of an activated electrically magnetized horse shoe magnet. A primitive brush system rubbed against the edge of the disk and the spindle. Faraday found that when he turned the generator in a clockwise direction he would have a current flow in one direction and then when the disk was turned in the opposite direction it would also reverse the direction of the current.

This device was given several names but homopolar generator is the one I have chosen to use. The original was very inefficient as the eddy currents produced counteracted the flow of electricity but the experiment was successful in showing that there was a relationship between magnetism and electrical generation and led to the fundamentals of all modern electric motors and generators.

Nikola Tesla looked at the Faraday homopolar generator and devised an improvement to it. He realized that the brushes were essential to pick up the electrical charges but knew that brushes wear down quickly and therefore are not very efficient. Others had realized that one way to replace the brushes was to create a bath of mercury, which being a liquid metal, allows the edge of the disk to be dipped in it creating a brush like effect without the friction of the brush. The electricity can then be collected from the mercury bath by a lead wire being dipped in the bath. Nikola, however, decided that a better method of collecting the charge was to use the perimeter of the disk as a flange in which he put a belt impregnated with metal to collect the charge from the two disks (see diagram below.) Nikola also knew that a phenomenon known as the Faraday paradox, stated that if the magnets were attached to the copper disc and rotated, an

Can Tesla Save The World ?

electrical charge would be generated (also known as the Lentz Law). Thus if the one set of magnets and copper disk is rotated in a clockwise direction and the other set is also rotated clockwise, if the magnets on one disk are facing North/South and the other disk is South/North, the result will be a charge on the perimeter of the North/South arrangement that will be positive and the charge on the perimeter of the South/North arrangement will be negative (using the Fleming right hand rule).

This is shown below as Tesla's U.S. patent # 406,968.- Dynamo Electric Machine

Can Tesla Save The World ?

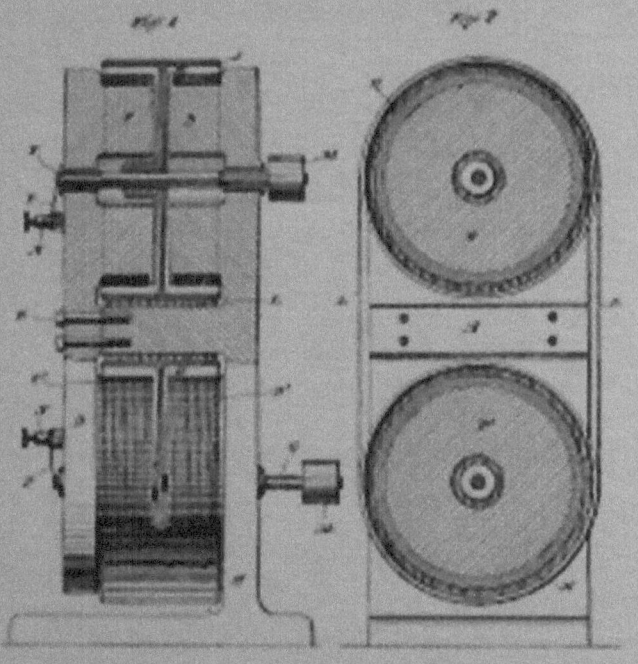

Can Tesla Save The World ?

In the proposed homopolar generator, the spindle of the disk can be larger than the original Faraday model with no difference to the output. The horseshoe magnet is replaced with stationary magnets that are created as shown in Fig. 1 below, which shows permanent magnets surrounded by iron disks that are wound with insulated copper wire. (The permanent magnets can also be surrounded by steel disks as these electromagnets will have a permanent polarity.) The copper wire surrounding the disks will be activated by the alternative energy sources of windmills and solar panels.

The stationary magnets will then induce magnetism into the stainless steel disks that abut the copper disk. Since the magnetism in the stationary electromagnets will be very powerful, this power will be conveyed to the stainless-steel disks and thus a good magnetic field will be present when the copper disk and stainless steel disks begin to turn. Since stainless steel retains its magnetism to a certain extent this will create a semi-permanent magnet either side of the copper disk. The size of the disks would be uniform but the diameter can be two to three feet if needed or more, as Nikola originally indicated.

As mentioned above Nikola Tesla decided to use a flexible, metal impregnated belt to connect the edges of the two disks. In the event that the belt should break then the inlet to those Tesla turbines will be shut off and the homopolar generator unit for that section will be removed. A spare may then be inserted in its place while the unit is being fixed. To fix the broken band the homopolar generator will be built to allow the two units of the machine to come closer together allowing the new band to be passed over the sections until it is in the position where it is supposed to be. The two units will then be pulled apart so that the tension on the band is restored and the unit is ready for use. Probably this homopolar generator unit will be stored until another band breaks and thus this becomes the replacement unit. The band (belt) will be made from any substance that will be pliable and yet be strong. One

Can Tesla Save The World ?

possibility may be graphene but this will have to be investigated.

One feature of the homopolar generator is that it generates low voltage but high currents and therefore is ideal for electrolysis; which is the purpose for this electricity. Although this direct current electricity will be supplemented with additional power from the windmills and solar panels, it is expected that the homopolar generators will provide most of the electrolysis energy for the air- electric system.

Besides the modification of the stationary magnets inducing magnetism into the moving magnets, another improvement will be that the spindles will be connected directly to the Tesla turbines. This will mean that two Tesla turbines will be used to turn spindles on both sides of the turbines as shown in Fig. 2 below. This also means that the two leads from one side of the generator will be duplicated on the other side, allowing the electricity to power two electrolysis systems at once.

The method by which the charge will be collected at the points of the spindles is illustrated below, in that the points connect with the terminals directly, and therefore will extract the electricity from the system.

Another way may be to create metal collars in which there will be steel ball bearings. This will be put onto the spindles and thus have a dual purpose of both supporting the spindle and also be a collector of the electricity that is on the spindle. More than one collar will be used and be connected in series to collect the electricity from each spindle.

(Again, as with other chapters, the Internet can give you a more detailed description of the operation of the homopolar generator also called the Tesla Unipolar Machine.)

Can Tesla Save The World ?

Tesla Homopolar Generator

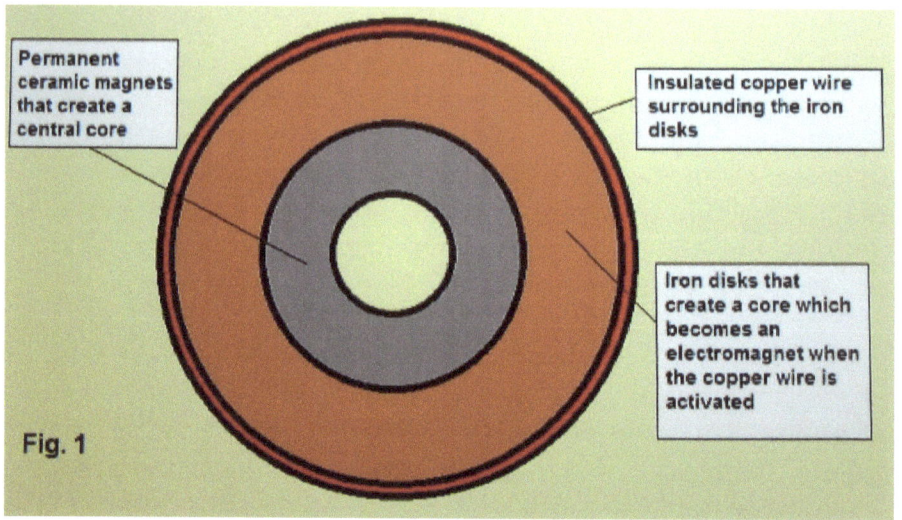

Permanent ceramic magnets that create a central core

Insulated copper wire surrounding the iron disks

Iron disks that create a core which becomes an electromagnet when the copper wire is activated

Fig. 1

Can Tesla Save The World ?

Tesla Homopolar Generator

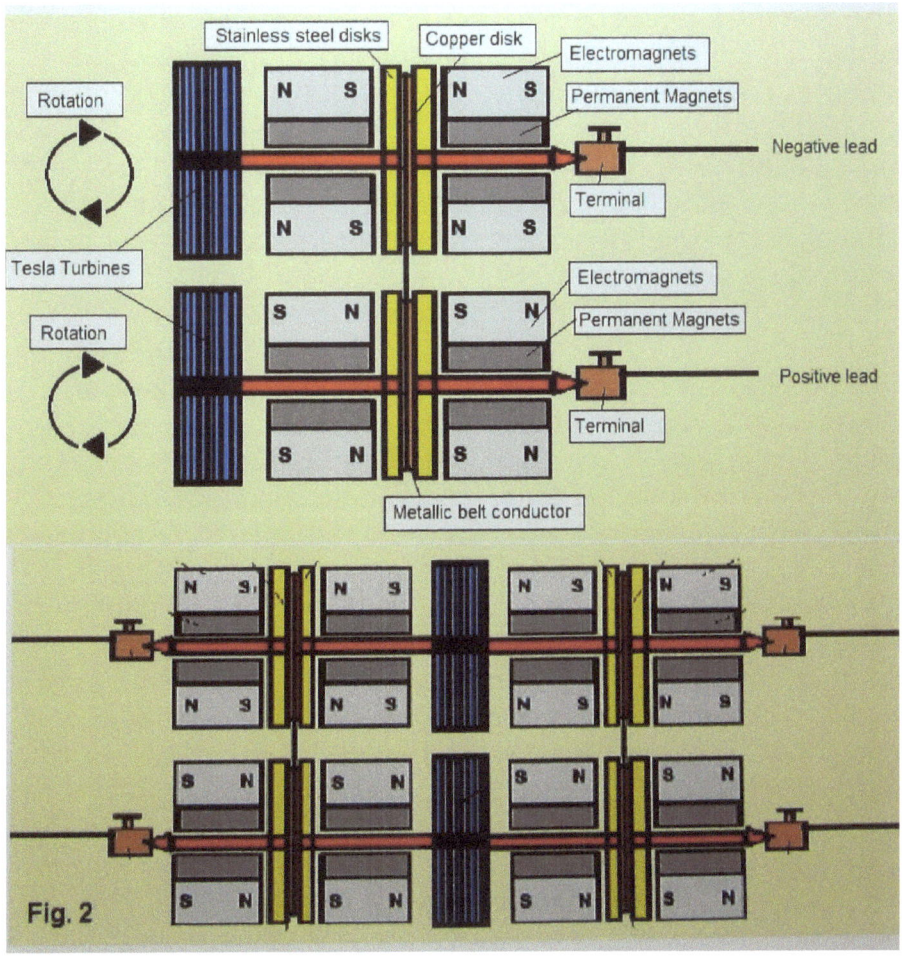

Fig. 2

Can Tesla Save The World ?
Oxyhydrogen Gas Generator

The oxyhydrogen generator was first used by William Nicholson in 1800 to decompose water. He discovered that if he split water by electrolysis into hydrogen and oxygen he created the formula: electrolysis: $2 H_2O \rightarrow 2 H_2 + O_2$ combustion: $2 H_2 + O_2 \rightarrow 2 H_2O$. He also discovered that it takes more energy to split the water than to burn it but as the mother plant is not relying on this operation to any great extent and that we are using alternative energy sources; this fact does not have any effect on the air electric system.

The apparatus I envisage, as described in Fig. 1, shows a positive plate with five neutral plates in between the negative plate. The length of each plate, that will be made of stainless steel, will be at least five feet long and twelve inches wide, which will create enough gas for the operation of the steam boiler. The oxyhydrogen gas generator will be connected to the terminals of the homopolar generator and as mentioned before, will be supplemented by the alternative energy sources. Water for the generator will be supplied by the outlet from the Tesla turbines, but since it is distilled water then potassium hydroxide will be added to the generator as a catalyst.

The resulting gas will be channeled through a bubbler to avoid back draft into the generator. As the pressure builds in the bubbler, the oxyhydrogen gas will be sent to the next stage: the steam boiler, where the gas will be exploded by glow plugs, powered by the batteries fed by the windmills and solar panels.

The level of water in the oxyhydrogen generator and in the steam boiler will be controlled by the level of the tank into which the water from the Tesla turbine will enter. Any excess will be diverted by a water pump (previously described) to other areas of the operation.

Can Tesla Save The World ?

The next section will describe how the oxyhydrogen gas generator will power the steam boiler.

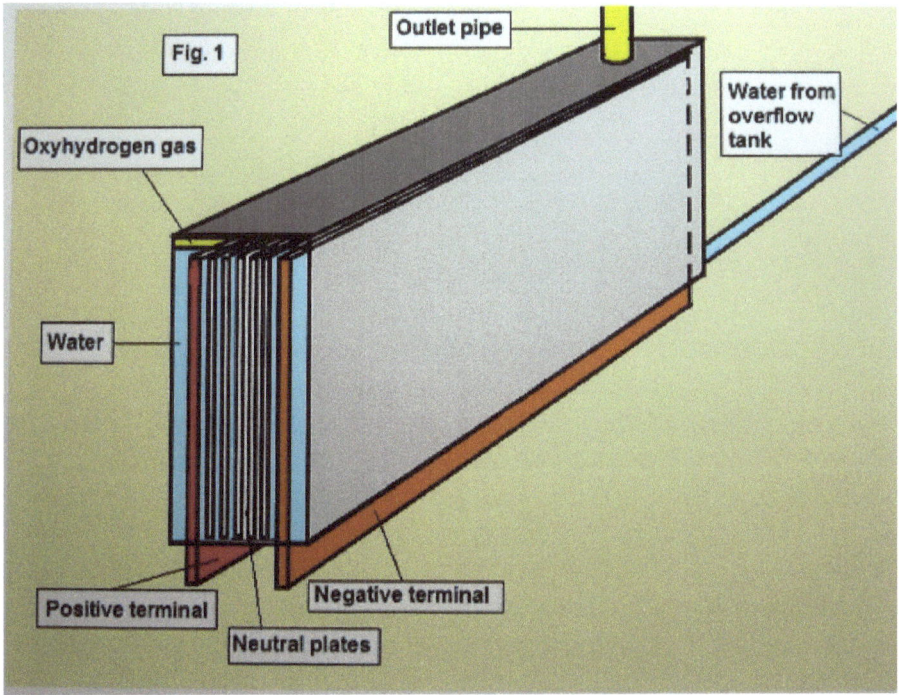

Can Tesla Save The World ?

Steam Boiler

The steam boiler will be the method for raising the water from the run-off of the Tesla turbines back to the original reservoir at the start of the trompe system. Thus, the steam boiler acts as a pump rather than any kind of engine.

Fig. 1 shows a diagram of one section of the steam boiler but as with all parts of the air electric system the sections will be expanded to meet the demand for the compressed air of the mother plant. The oxyhydrogen gas from the bubbler enters the base of the steam boiler and flows through the water into a small bell shaped dome. The pressure of the gas will push the water out of the dome and as the capacitor reaches its maximum charge the glow plug will ignite the gas and cause a mini explosion. The oxyhydrogen will then release its energy in the form of heat and the gas will revert to water. This heat will then effect the surrounding water so that the surface water will eventually boil. (Note that there will be many bell domes and thus quite a bit of heat.) The surface water will then boil to the point where steam is created and as the steam pressure builds it will be channeled out of the boiler via thin tubes that will rise vertically to a height of an hundred feet or more. (Note that at some point the bell domes will become hot enough to sustain the explosion of the oxyhydrogen gas and thus at such a point the glow plugs will be shut off.)

These exit pipes will be clustered and have insulated copper wire wound around certain sections. When direct current is passed through this wire it will heat the pipes and therefore help to maintain the steam inside them.

When the pipes reach the top of their one hundred feet they will traverse the length of the trompes but be above them, shrouded in earth. The earth will provide a means for cooling the pipes and thus condense the steam in them back to water. This water will flow back into the trompe reservoir to allow the process to begin again. (See overview at the end of this

Can Tesla Save The World ?

section.) The earth will be at least four feet deep or below the frost line of the area in which the mother plant is situated. This means that the mother plant will be able to function in all seasons, including winter.

An alternative cooling method would be to have a system of returning some of the water from the overflow of the first reservoir to a height above the pipes carrying the steam. This will mean that they will aid in condensing the steam back to water as well as adding water to the first reservoir. To accomplish this will mean to use a pump known as a hydraulic ram pump, where the water runs down a pipe and out of a flapper valve. As pressure builds up, the flapper valve closes and forces the water back to another one-way valve. As this opens. the water rushes into a container and compresses the air in the cylinder. The one-way valve then closes and the compressed air pushes the water up another pipe. The result is that the water may be lifted to quite a height but as most of the initial water escapes through the flapper valve only about ten to fifteen per cent of the water flows up the escape pipe. Inefficient as this is, this method will supply enough water to help condense the steam and since this pump is not using any external energy to operate it, it will be effective enough for our use. Fig. 2 shows a typical hydraulic ram pump set-up but for a better explanation you can look on the internet or watch a You Tube presentation on its operation.

(You Tube – **How Ram Pumps Work**

https://www.youtube.com/watch?v=aUTjVovpKvA

Can Tesla Save The World ?

Steam Boiler

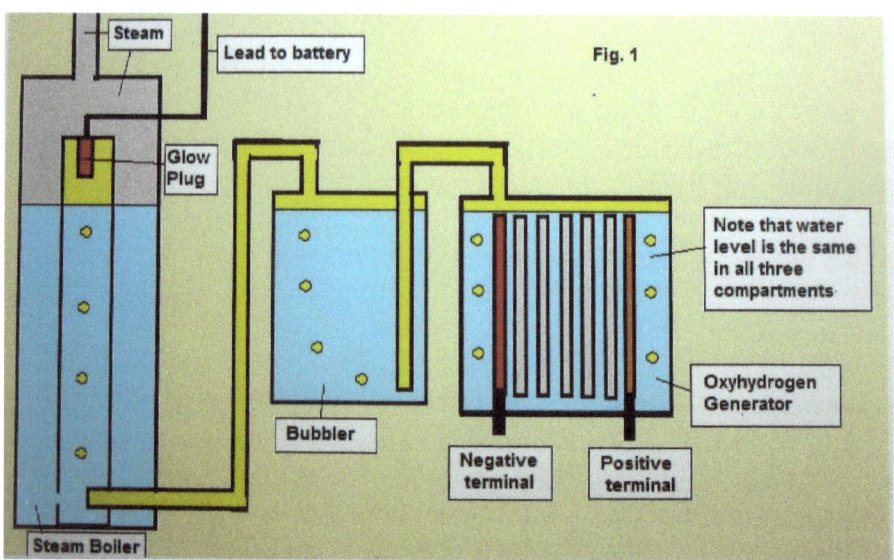

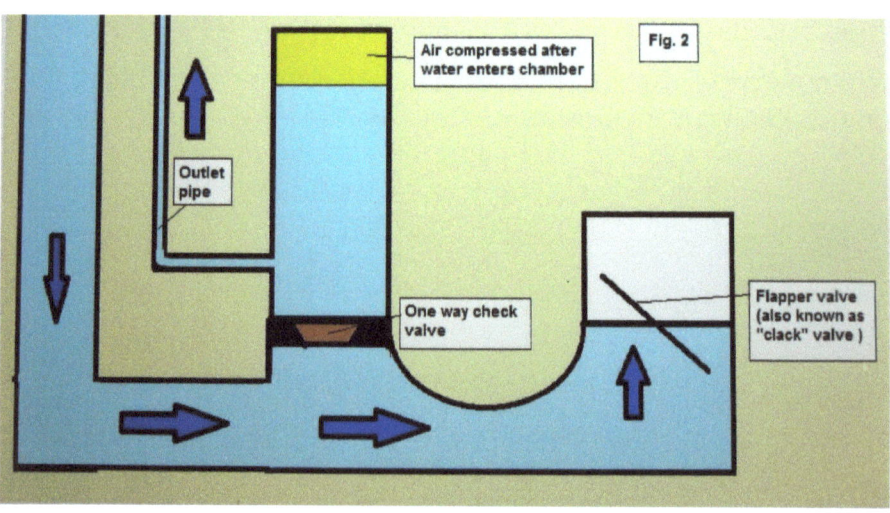

Can Tesla Save The World ?

Gasometers

Historically gasometers have been used for storing natural or coal gas. They have been built up to 100 feet high and although they are rarely seen nowadays, their function to hold large volumes of gas under pressure was impressive.

The air electric system will utilize gasometers to allow the compressed air from the trompes to be stored until the air is passed on to the satellite plants that will generate the electricity to be used locally.

Fig. 1 illustrates how these gasometers will work. Essentially a block of trompes will allow the compressed air they collect to enter the gasometer. As the dome rises due to the compressed air pressure inside, a point will be reached where the air is to be sent to the satellite stations. Just like the water pump previously described, the gasometer will have a fixed tank above it that will be full of water. When this water is released into the movable tank, attached to the gasometer, the weight will cause the dome to further compress the air within so that the pressure of air leaving the gasometer will be quite substantial. As the dome descends under the weight of the water, there will be pistons to the side of the gasometer that will force water up tubes and will fill the fixed tank to replace the water that has been used in the process. (One-way check valves will stop water from leaving the inside of the gasometer.) In turn the water in the movable tank will exit when the dome reaches a certain level and fill the space left by the water pumped by the pistons. Therefore, the gasometer will always maintain a relative level of water within it whilst delivering substantial amounts of compressed air to the satellite stations. (Any water lost to evaporation will be replaced from the trompe refill system.)

Fig. 2 illustrates the inside workings of the gasometer. The compressed air from the trompes is fed to the gasometer until the upper dome is pushed up to where the rubber seal on the

Can Tesla Save The World ?

edge is at its maximum. (There will be a rubber seal on the movable part of the gasometer as well as another at the top of the fixed part. This will prevent water or compressed air escaping.) When the maximum height is reached the one-way-check valves will be activated so that both the compressed air coming in and the water will be shut off. At this point the water will enter the tank attached to the gasometer and the weight will force the dome to descend pushing the compressed air in the gasometer into the outlet funnel, which is connected to the piping that takes the compressed air to the tunnel system and then on to the satellite stations.

The gasometers will fill up at different times and so their contents will enter the tunnel system at different times also. This will mean that there will always be a consistent supply of compressed air which will be diverted to the various destinations, whether satellite stations or rural/business locations, and as the gasometer would have increased the psi of the compressed air it will move along the pipes at reasonable speeds.

The one-way-check valves are shown in more detail in Fig. 3 and Fig. 4. The valve utilizes electromagnetism to pull the plunger one way, thus opening the valve and then the other electromagnetic coil will be activated to close the valve. Although the plunger of the stopper will be made of iron it will sheathed in plastic to avoid the chance of rust. The effect of the valves will be that the gasometer will be completely tight allowing the lowering of the dome to further compress the air before it is sent to the tunnel system.

Can Tesla Save The World ?

Gasometers

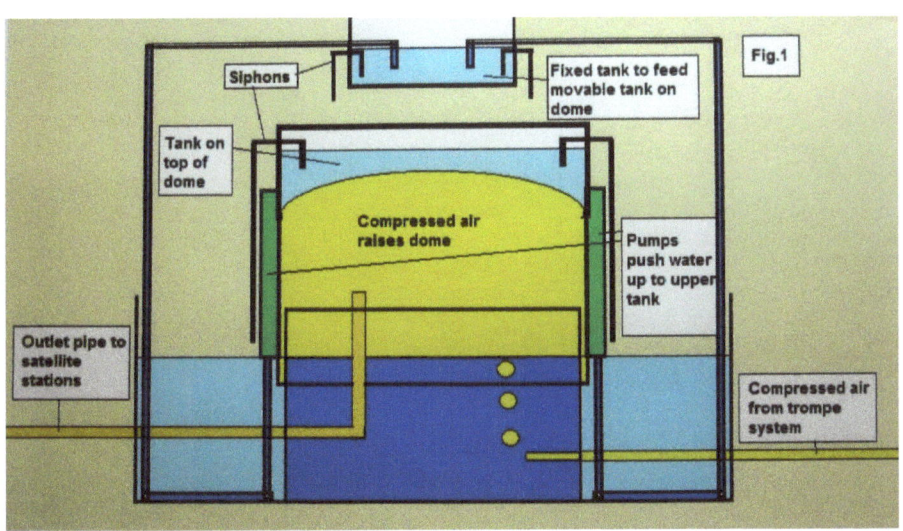

Fig.1

Siphons

Fixed tank to feed movable tank on dome

Tank on top of dome

Compressed air raises dome

Pumps push water up to upper tank

Outlet pipe to satellite stations

Compressed air from trompe system

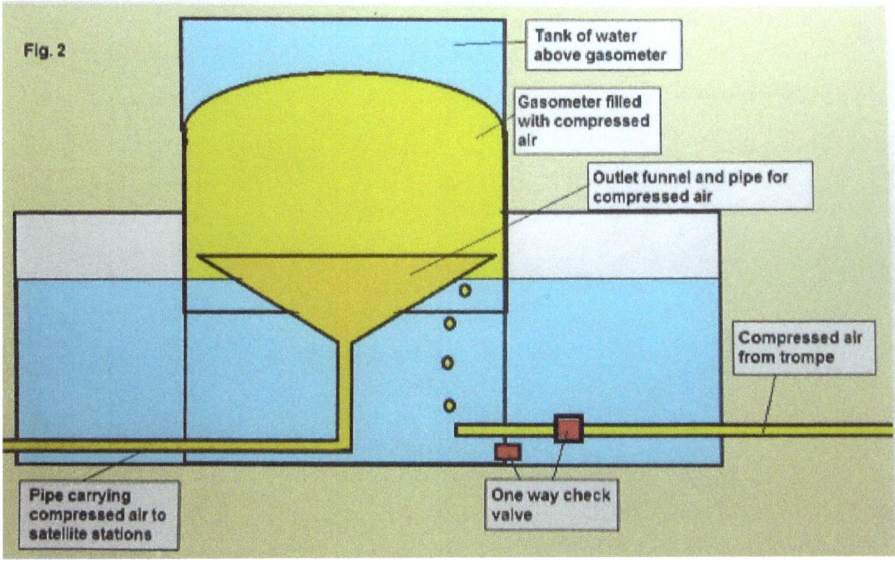

Fig. 2

Tank of water above gasometer

Gasometer filled with compressed air

Outlet funnel and pipe for compressed air

Compressed air from trompe

Pipe carrying compressed air to satellite stations

One way check valve

Can Tesla Save The World ?

Gasometers

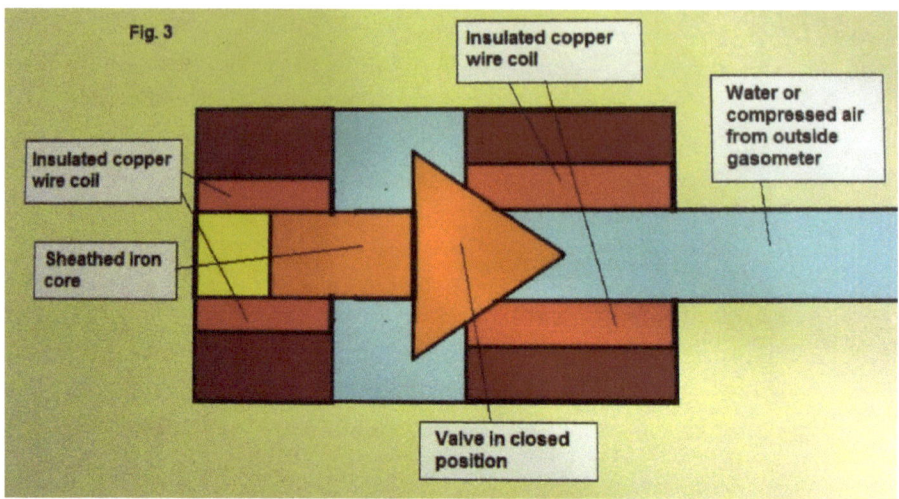

Fig. 3

Insulated copper wire coil

Insulated copper wire coil

Sheathed iron core

Water or compressed air from outside gasometer

Valve in closed position

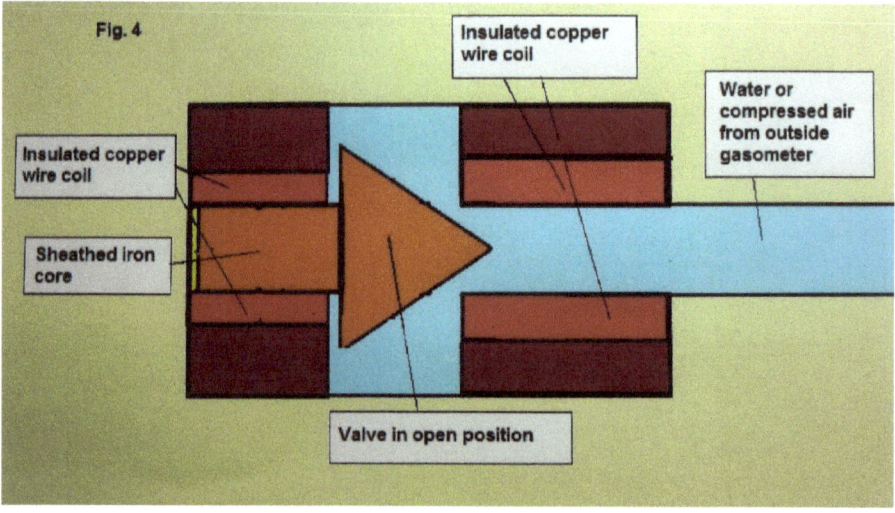

Fig. 4

Insulated copper wire coil

Insulated copper wire coil

Sheathed iron core

Water or compressed air from outside gasometer

Valve in open position

Can Tesla Save The World ?

Mother Plant Overview

As can be seen from the descriptions above, the only purpose of the mother plant is to produce compressed air that is fed to satellite stations. The water in the trompe system is therefore recycled through the use of the steam boilers; raising the water in the form of steam and then releasing it as water through condensation. The compressed air collected from the base of the trompes fills gasometers, that in turn feed the satellite stations.

The mother plants will be situated near to rivers, lakes or areas close to water access. They will be initially fed by these supplies of water but once the system is filled the mother plant will only require additional water due to evaporation. This can probably be supplied by local rainfall or snow. Being below ground they will not be subject to either heat or cold and therefore be able to operate all year-round. Since wind patterns are usually prevalent in areas close to bodies of water, the use of windmills to generate supplemental electricity should not be a problem and since the mother plant can be in a remote area, noise pollution or other problems related to windmills will not be an issue. The use of solar panels will also cause no problem, again because of the remoteness. Since compressed air can travel large distances with very little loss of pressure, the stainless steel pipes that conduct the compressed air will mean that the system will not rely upon power lines and the resulting towers associated with the present grid system. Instead the stainless steel piping will pass underground, possibly in tunnels similar to sewers, to reach the satellite stations. This should also mean that the air electric system will not be effected by ice storms that presently create havoc to the overhead power lines.

From an environmental aspect the fact that after the initial water is used very little extra will be needed and since the mother plant will create isothermal compressed air, when it is released to the atmosphere it will be receiving better air than

Can Tesla Save The World ?

was taken from it. The initial water and additional water will be filtered such that it will be very clean and then having passed through the steam boiler will become distilled and thus become even cleaner. This will mean that periodically the steam boiler will have to be cleaned but probably not for many years.

The last major feature of the mother plant is that there is no end to its limit as the end result of compressed air means that to extend its supply only means to add more trompes to the system. It is envisaged that twelve trompes (four sections of three trompes) be used as a standard unit but there is no end to the number of units that can be operating.

(See the air electric layout below for a schematic of the mother plant. The initial feed of water and subsequent fill-up water is not shown. Also the water pumps are not shown as the description of their use in the Tesla turbine over flow reservoir and the gasometer have been described above. The diagram is also not to scale but is intended to give a visual of the mother plant in its simplicity.)

The layout of the exterior of the mother plant is shown in Fig. 1 where a pyramid type structure is created to contain the trompes and steam boilers. The pyramid is expected to rise 90 feet in the air allowing for at least another 90 feet to be dug out to accommodate the 150 foot trompe pipes. As can be seen the gasometers will be situated at ground level but deliver their contents to the tunnel system below. The soil-slots in the different steps of the pyramid will be filled with earth and therefore, will allow the ability to raise crops of some kind. However, the earth at the top of the pyramid is to support the solar panels and windmills as well as acting as a condenser for the steam coming from the steam boilers.

Can Tesla Save The World ?

Mother Plant Overview

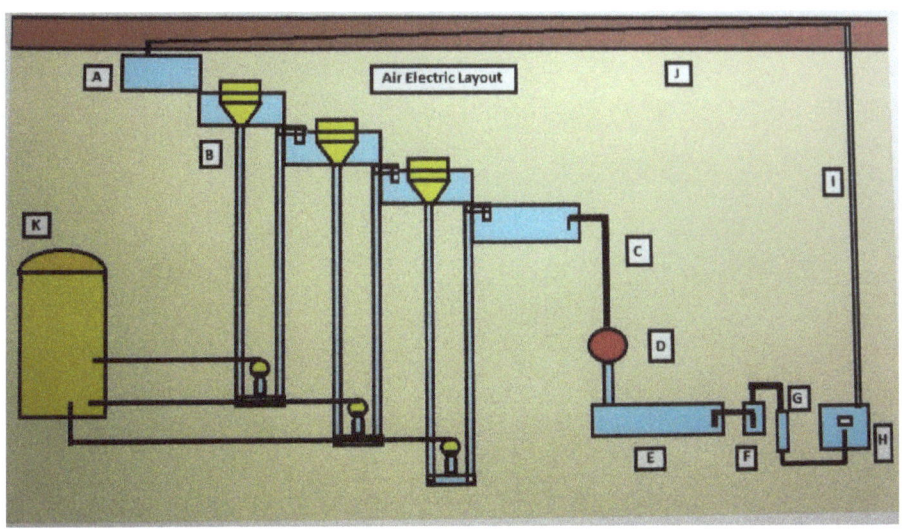

Air Electric Layout Legend
A - Reservoir for trompes   B – Trompes   C – Overflow Siphon
D – Tesla Turbine and Homopolar Generators   E – Reservoir for Tesla Turbine   F – Oxyhydrogen Generator   G – Bubbler
H – Steam Boiler   I – Steam pipes   J – Earth to condense steam K- Gasometer for compressed air from trompes

Can Tesla Save The World ?

Mother Plant Overview

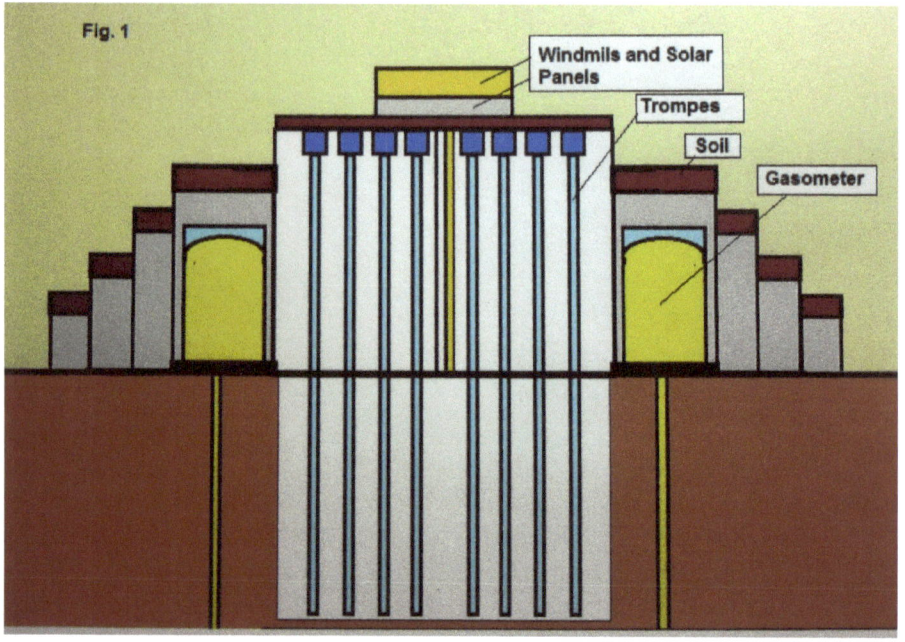

Can Tesla Save The World ?

Satellite Stations

As mentioned earlier the compressed air will pass from the mother plant to satellite stations. These may be many miles away and yet the compressed air will have no problem reaching them. The compressed air will enter a gasometer similar to the one in the mother plant and this reserve will operate the satellite station.

There is a pump known as the Brumby Pump that originated from Australia. It is basically an air lift pump but uses a ball arrangement to make a very efficient pump that can lift water many feet into the air. It is simple to operate and uses only compressed air to run it, (no electricity needed). The manufacturer guarantees that his pump will be almost maintenance free and that it requires little or no experience to operate. This pump will be used as the major pumping device for the satellite station.

The trompes in the satellite stations are identical to those in the mother plant with the exception that the outflow at the end of the trompe series will be one to two hundred feet and the drop will be fed to either Tesla turbines or Frances turbines that will be connected to the electrical generators. The turbines will run at 3,600 rpm in the North and South American systems and 3,000 rpm in European systems so that their 60 cps and 50 cps can be maintained. To accomplish this the outflow will use siphons that will be geared to the exact amount of water to create the turbine speeds needed.

This will also mean that the water being pumped to the trompe reservoir will also have to be accurate and thus siphons will also be used there to accomplish the correct flow. As has been mentioned before the trompe units in the satellite stations will also create compressed air along with the electrical output. This compressed air will be passed on to other satellite stations so that there can be a relay of satellite

Can Tesla Save The World ?

stations each capable of supporting the next for compressed air.

The initial water for the satellite station will be either from local sources, filtered and passed through a steam boiler to distill it or else the mother plant may supply it. Again once the initial water is obtained, any addition supply can be obtained from local rainfall, snow or local sources; such as streams, rivers or lakes.

Fig. 1 shows the setup of the of the satellite station, showing that the original reservoir is below the turbine level and will be filled by the runoff from the turbines. The Brumby pumps will then lift this water back to the top of the trompe reservoir and the cycle will be complete. The compressed air then used by the Brumby pumps will be released to the atmosphere and the new supply of compressed air will come from the mother plant (or other satellite stations).

The advantages to this system are that besides being totally environmentally friendly the satellite stations can be set up either in the towns, cities or local areas without any fear of pollution. The wires from the satellite station will look like any electrical generating plant with the exception that as only local electricity is being provided there will be no need for the high voltage power lines.

In rural areas it may be practical to set up individual electrical stations for each home or farm, thus the compressed air will be used to run Brumby pumps but there would not be a need for trompes and thus a circulatory system as shown in Fig. 2 would suffice. This method may also be used for individual businesses in towns that need an independent electricity supply.

The satellite stations would therefore be able to meet any electrical needs of any community in a clean and efficient manner. Storms that presently effect local power stations would still pose a threat to transformers, etc. but the repairs

Can Tesla Save The World ?

should be easier to perform than having to worry about ice storms or high winds effecting power lines. Thus power outages will be minimized and in most cases be localized.

Satellite Stations

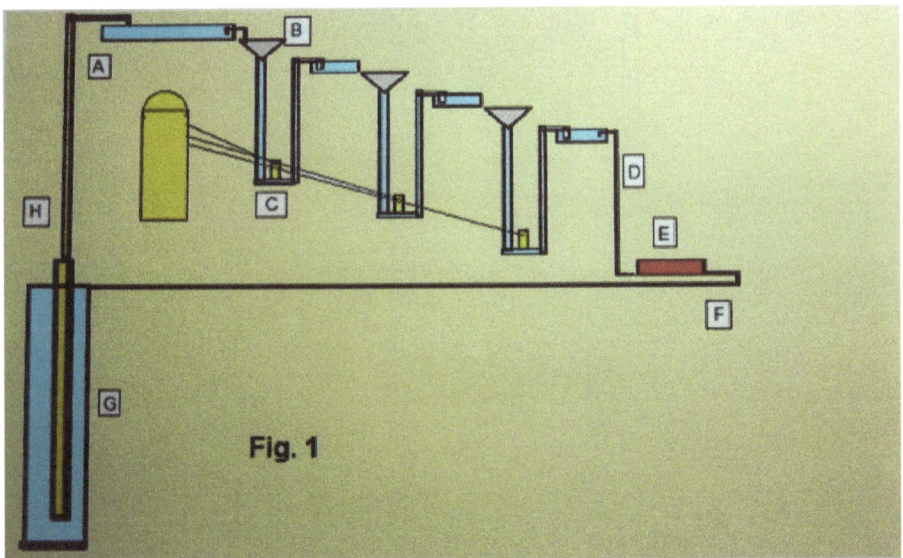

Fig. 1

A – Reservoir B - Trompes C – Compressed Air that fills the gasometer D – 100 foot drop of water E - Electrical Generators F – Water returns to tank (G) G – 300-foot tank with Brumby pump (H) which raises the water up 150 feet to fill reservoir A to complete the cycle.

Can Tesla Save The World ?

Satellite Stations

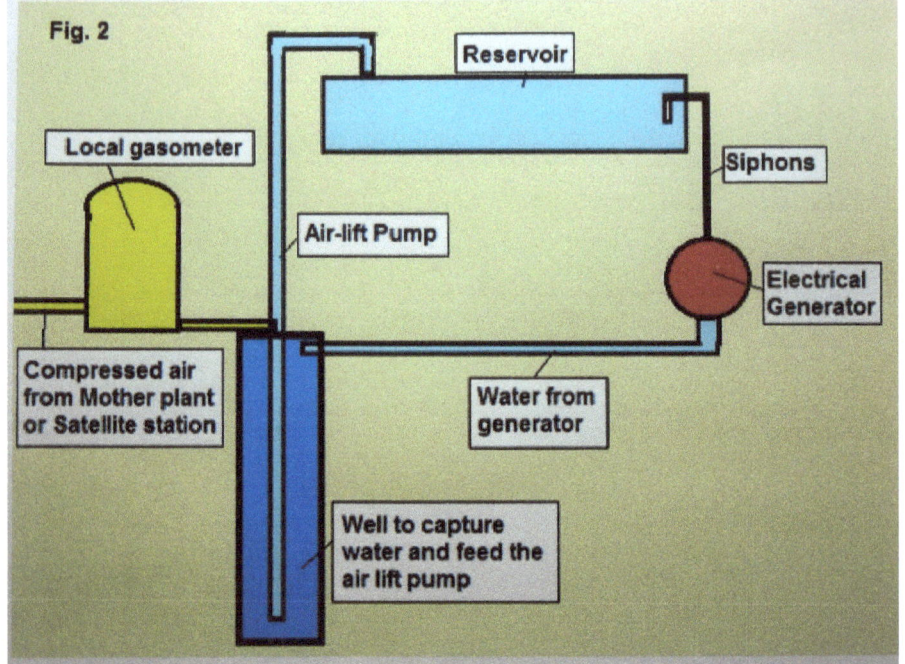

Fig. 2

Local gasometer

Reservoir

Siphons

Air-lift Pump

Electrical Generator

Compressed air from Mother plant or Satellite station

Water from generator

Well to capture water and feed the air lift pump

Can Tesla Save The World ?

Distribution System

The compressed air from the mother plant and any from the satellite stations will be distributed, like the Viktor Popp idea, by using tunnels. Popp used the Paris sewers as his distribution system but as the mother plants will be in remote areas it is expected that pipes will be laid to carry the stainless-steel piping for the compressed air. However, as these pipes will transverse large distances and probably through private property it may be practical to have these tunnels at a depth of about sixty feet thus not interfering with the land above it. If the piping is to go into lakes or rivers it may be practical to either go around them or underneath them but if the lake or river is too deep, then the tunneling may be built to go along the lake-bed or river-bed.

It is envisioned that the tunneling will be at least seven or eight feet in height as it will need to allow a person to inspect the stainless-steel piping periodically. This may mean that a vehicle be used for this task but it will also mean that there will be intermittent air passages from the surface to allow the person to breathe.

It may be possible that the local sewers may also be used as a conduit for the stainless-steel piping but this will be worked out with those municipalities involved.

What this does mean is that as electricity is not being carried long distances as there will no need for power-lines that presently carry the high voltage wires. Instead there will only be a need for local utilities to provide poles to carry wires from one transformer to another and then to the user.

In the case of rural areas or when the compressed air is being piped directly to a generating station situated on the grounds of a factory or business, the station may look like a satellite station but without the use of any trompe. The distilled water will then be shipped to these generating stations, probably by

Can Tesla Save The World ?

water trucks, and be used to top up the system after the initial filling-up process has been completed.

Can Tesla Save The World ?
Wind and Solar Energy Sources

As mentioned earlier, the supplemental power for the air electric system will come from wind and solar power. On top of the pyramid roof of the mother plant will be two displays of renewable energy. The solar panels will be situated on a large tarmac area such that servo-motors will point them towards available sunshine. Since this technology is not new it will suffice to say that the solar array will pick up the solar energy and pass it to batteries, (described later in this chapter.) The tarmac will be tilted so that besides supporting the solar panels it will also be a collector for rainfall and winter snow. This will be fed to storage tanks to be used as top up liquid to the overall air electric water system.

The wind power will also be collected by batteries but the wind turbines will adopt the more versatile Savonius windmills. The advantage of these windmills is that they are multi-directional, in that regardless of the direction of the wind, the Savonius windmill will be able to utilize it. The pyramid structure of the Mother plant will be at a height of ninety feet and thus the Savonius windmill will not need large poles to reach the wind, allowing the electricity generated by the windmills to be easily passed to a battery system. Fig. 1 shows the top of the pyramid arrangement and the location of the solar panels and windmills. Fig. 2 illustrates the envisioned Savonius windmill arrangement to be used in this project. Note that the permanent magnets are directly connected to the windmill and therefore will generate direct current electricity without the use of gears. If the prevailing winds are too severe a barrier can be raised to prevent the windmills getting out of control, thus eliminating the need for a braking system. As the main continuous-wire will collect the electricity from each electromagnet, the total output will be quite substantial, even at lower wind speeds.

Fig. 3 illustrates the Savonius windmill from a side view. This is only portion of the complete windmill and shows how the

Can Tesla Save The World ?

eight sections of the Savonius windmill are situated and thus will provide the necessary torque. The permanent magnets in the central section are fixed so that as they spin they are able to induce a current in the electromagnet which is then transferred to the main wire. As the main wire will collect all the electricity from the different electromagnets the permanent magnets will all face the same direction (either all North/South or all South/North). Although the windmill will pick up any wind direction it will only spin in one direction and thus the electricity produced will be consistent.

The battery system will probably use the zinc-bromine batteries; as they have the capacity to store the electrical output of both the solar and wind energy and can be charged and discharged many times with no ill effect on the battery. There has been some remarkable research into this type of battery using a gel instead of a liquid electrolyte in Australia. Inverters may also be used to transmit electricity to different parts of the mother plant using alternating current.

Windmills and Solar Panel Energy Sources

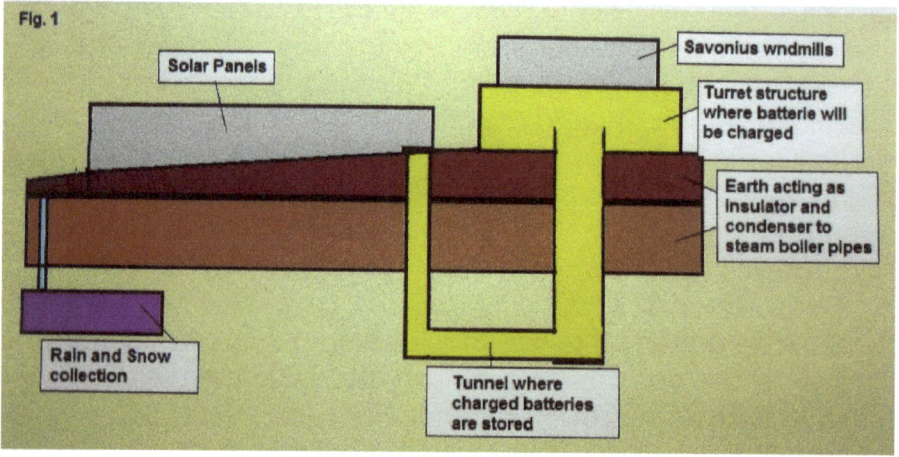

Fig. 1

Solar Panels

Savonius wndmills

Turret structure where batterie will be charged

Earth acting as insulator and condenser to steam boiler pipes

Rain and Snow collection

Tunnel where charged batteries are stored

Can Tesla Save The World ?

Wind and Solar Panel Energy Sources

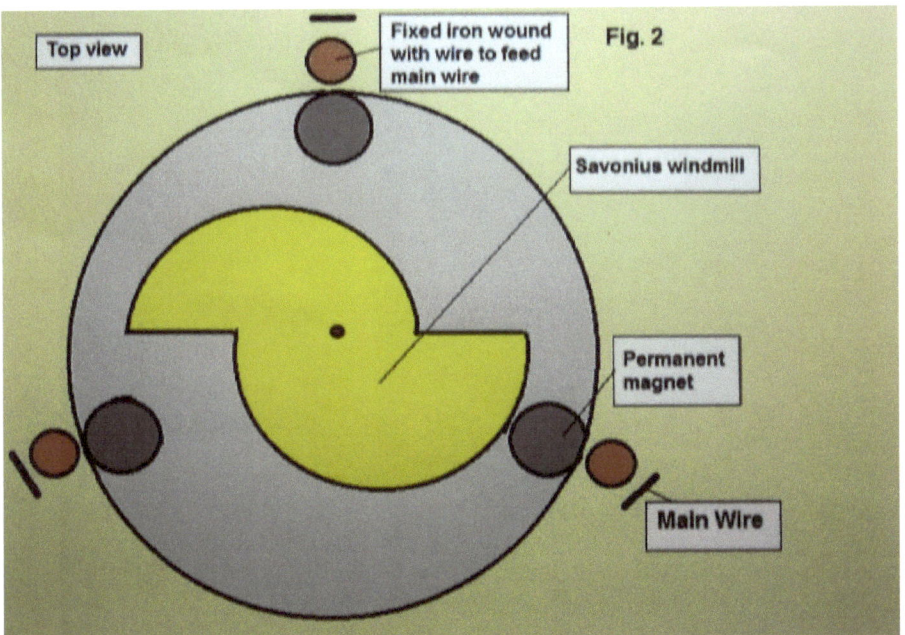

Can Tesla Save The World ?

Wind and Solar Panel Energy Sources

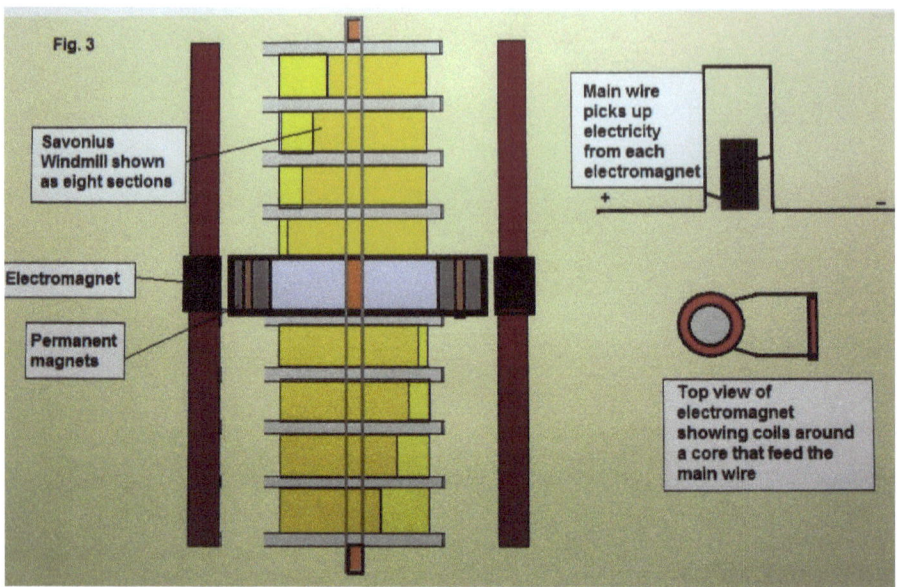

(The windmill shown here is only one section but the working windmill is expected to have several sections creating a Savonius windmill that will be about 8 feet high and 30 feet long.)

Can Tesla Save The World ?

Overview

Thus hopefully the electrical system utilizing the air electric system has been explained sufficiently for any utility to build such. The mother plants will supply compressed air to satellite stations, which in turn will generate local electricity and also have the capability of passing on more compressed air. The air electric system will be adaptable to any area, however large, and being that its operations are mainly below ground are not subject to the heat or cold of the seasons.

Although a water supply will be needed at first, the mother plant and satellite stations will only need topping up after the start of their operations, which should be supplied by either rainfall or snow. Should it be necessary for the mother plant to supply water to a satellite station then a constant supply of water will be needed but as this can be a lake, river or even sea water, (though it would need to be distilled), the supply should always be available.

Going back to the beginning of this explanation the major advantage of this system is that it is environmentally friendly and thus should help to eliminate present day pollution from power plants that effect the climate. Should this method be adopted it is hoped that our children and grandchildren will have a brighter future and still be able to meet an ever increasing demand for electricity.